AF760129

BEYOND the CLOUDS

BEYOND the CLOUDS

Chip Ricks

Tyndale House
Publishers, Inc.
Wheaton, Illinois

Credit: Cover photo and several others used in this book were taken by John E. Granado.

Library of Congress Catalog Card Number 79-65024
ISBN 0-8423-0149-8, cloth
Copyright © 1979 by Chip Ricks
All rights reserved
First printing, October 1979
Printed in the United States of America

In memory of Connie's granddad,
who knew that there is love and
life beyond the clouds.
September 14, 1898–October 29, 1978

CONTENTS

FOREWORD

Connie Engel is a jet pilot in the United States Air Force. She is also a pioneer in military social change.

A generation ago, during the difficult days of World War II, American women first flew U.S. military aircraft as members of the Women's Air Service Pilots (WASPS). They made a great contribution to the war effort and, though not exposed to combat, proved that women could fly the front line aircraft of that time. They logged thousands of hours in the air, coped expertly with emergencies and bad weather, and exercised the good judgment and firm leadership the flying mission demands. They served their country well.

At the end of the war, the WASPS were disbanded. Over thirty years later, the United States again has women serving as pilots and navigators in military cockpits — now in high-performance jet aircraft. Connie Engel is among them.

A central reason for the change is the realization that women who wish to pursue a military career should be able to apply their skills as fully and equitably as possible. The Air Force has steadily expanded career opportunities for women, and they now perform duties in 98 percent of the fields open to men. The program to include women in pilot and navigator fields is a logical continuation of a basic policy.

Connie Engel has helped to prove the merit of that policy. She had a distinguished record as an Air Force junior officer and nurse before she applied to be a member of the first group of women to enter Air Force pilot training in 1976. She was one of eighteen women selected from among seventy very talented and eager applicants. Ten were assigned to class 77-08.

A year later Connie graduated at the top of her class—competing against men and women alike—a standing she earned by exceptional flying skill and leadership. In winning her wings, and the coveted Commanders Trophy given to each class's outstanding graduate, she became "first of the first."

Connie grew up immersed in Air Force tradition. Her father was an Air Force officer for thirty years, and spent much of his career as a bomber pilot in the Strategic Air Command. She understands the fundamental values of duty and country, and has a marvelous zest for life. She has carved her place in what has long been a man's world, and has excelled.

She exemplifies the talent and high spirit that abound in the young people today who have chosen to serve their country in uniform. I am proud of her — and of them.

Lew Allen, Jr.
General, USAF

FOREWORD

Captain Connie Engel was one of the first ten women to become jet pilots in the U. S. Air Force. She and her companions were the vanguard of a courageous group that proved women can meet the grueling physical and mental demands of high-speed, high-altitude flight. They were the pacesetters, the first to tread in what had been exclusively a man's world. The challenges they faced were enormous; their achievements made history.

Beyond the Clouds reveals the rich human drama Captain Engel experienced before, during, and after flight training. We see her both as a woman, deeply moved by the people and events around her, and as a proud member of the Air Force, determined to make the world a better place. We feel her frustrations and pain, and we come to know her intimately. We learn the source of her inner strength, and see how she shared that strength with others. And in the end we join in her triumph, a product of intense personal dedication, talent, and an unwavering faith in God.

The Air Force is deeply indebted to Connie Engel and others like her. Thanks to them, and the challenges they met, the sky has opened wide for women. The Air Force and the nation are stronger because of them.

John W. Roberts
General, USAF (Retired)

PREFACE

This is the story of a young woman's inner struggles as she dealt with such problems as the relationship of life to death, her purpose as a woman in a predominantly man's world, and her understanding of what it means to "walk in truth."

I have worked closely with Connie and Rich Engel to bring you a story which is factually true. In all but a few instances real names of people and places have been used. At the same time I have taken the liberty of fictionalizing a few conversations and compressing some incidents into a shorter time span so that the truth of Connie's search for answers to the big questions might retain its rightful place.

Chip Ricks

ONE
JOY IS DARKENED

Connie shifted her weight to her left foot and adjusted the phone on her shoulder as she continued to draw airplanes on a scratch pad. "Are you sure there won't be some slipup?" she asked. "Can you guarantee that I will be assigned to Williams Air Force Base in Arizona? It's not that I don't trust you, but I'd feel better if the promise was in writing—over an official signature, preferably some general's."

Connie glanced up at her mother, who had just entered the room and was listening intently to the telephone conversation. "O.K. Send me the papers. My dad will swear me in. Thanks for getting me assigned to Williams."

Hanging up the phone, Connie turned around, "Mom, how long does it take to get mail from Santa Barbara to Lompoc?"

"Never mind. You just tell me what that phone conversation was all about."

"I'm joining the Air Force."

"The Air Force? Just fifteen minutes ago I heard you tell the Navy recruiter you'd join the Navy."

"He offered me Pensacola, and that's no competition for Williams. Do you know that the sun shines 85 percent of the time in Phoenix? Besides, Williams is the largest undergraduate pilot training base in the Air Force. There are 300 men going through pilot training there right now. I'm twenty-three, Mom. And single."

"Yes, and you're very pretty. I know. Three hundred men, you say."

They both laughed.

"But I'll admit I like the idea of your being nearer home, Connie. And your dad will be pleased with your choice. Twenty-nine years in the Air Force has made him a little prejudiced."

Two days later the papers arrived from the Air Force Recruiting Office in Santa Barbara. Connie studied the small print and agreed to accept the terms even without a general's signature.

During the few weeks she had been home, she had explored every dress shop in Lompoc, Santa Maria, and Santa Barbara. Her closet bulged with clothes. But as she stood looking them over she thought, I have nothing to wear for a swearing-in ceremony. Holding up a pink flowered dress, she fingered the material. No, it's too fluffy. Besides, Dad won't like it; it's too short. I'll save it until I get to Williams. She replaced the dress on the rod and chose a tailored blue knit. Only Dad and Mom are waiting downstairs, she thought, and this one will please them.

When Connie came down the stairs, her dad was standing in front of the fireplace, one hand on the mantel, looking at the slow-burning logs. Tall and handsome in his blue uniform, he turned and watched his daughter enter the room. Connie noted that his hair was getting thin on top.

Her mom, whose short blonde hair blended with her beige dress, laid down the book she was reading and rose from her chair.

"You stand here," Dad directed her, "and Connie, you face me and raise your right hand."

The oath of allegiance was short, but Connie could see her dad had something more to say. He stood looking at her, studying her face, but the words didn't come. Connie put both arms around his neck, and he held her close.

"Lord," she heard him say, "take care of our Connie and stay close to her. Don't ever let her be alone. Teach her to depend on you."

Don't worry about me, Dad, Connie thought. I can handle the Air Force.

Early in November Connie took up her duties as a nurse on Ward 2 at Williams Air Force Base. The Air Force recruiting officer had kept his part of the bargain.

The young flyers at Willie did not disappoint her either. John Brown, who dropped by the hospital for a routine visit with the flight surgeon, introduced her to a group of junior officers at the Officers Club bar on her first day of duty. From that point on, she knew her decision to join the Air Force was a good one.

But the pilots at Willie soon learned that they were not the only men in Connie's life. T-38 jets from Reese Air Force Base in Lubbock, Texas, landed at Willie with enough frequency that some wondered if Arizona really was the new nurse's first Air Force assignment. One captain in particular, Jim Alexander, was often seen hurrying between the flight line and the base hospital. Jim had been largely responsible for Connie's decision to enter the service. She had been working at Saint Mary's of the Plains Hospital in Lubbock at the time, following her graduation from the University of Texas in June 1971.

An Air Force brat himself, Jim couldn't understand why anyone wouldn't jump at the chance to don the blues. "Besides," he argued, "you like to travel and the pay is good."

She had packed her bags in late August, said good-bye to friends, and headed for her home in California to think it over. The decision had followed a few weeks later after much research into all branches of the military service.

By early spring, after five months at Willie, Connie blended easily into the Air Force jet set. She had become accustomed to hearing the T-37 and T-38 jets streaking across the sun-drenched Arizona sky and to evaluating vivid and exciting reports of the pilots who flew them. On rare occasions these silver bullets claimed the lives of their occupants, but Connie had never seen even one victim of an aircraft accident, major or minor. She never thought about such things.

Now, coming off the seven-to-three shift at the hospital, she walked through the double doors of the gray frame building and down the steps to the sidewalk. Just then she saw Mike Conner, a handsome instructor pilot from Reese, coming toward her. She waved and knew he was using the seconds between them to look her over. She felt confident. A slender 114 pounds in her white nurse's uniform, her light brown hair piled in a bun on the back of her head, she slung her purse over her shoulder as she walked.

"Mike! What are you doing at Williams? Jim isn't here with you, is he? He said he'd be in at four tomorrow." The words tumbled out as Mike stood holding both her hands. Looking into his face, Connie sensed an uneasiness about him. He's probably tired, she thought.

"Hey! Hold the questions. I'll answer the first one," Mike said, squinting his eyes in the bright sun as the hot wind rumpled his sandy hair and tossed a few strands across his forehead. "I'm here on cross-country—have a student with me waiting on the flight line. Where can we talk?"

"How about my quarters?" Connie asked. "I've got a quart of sun tea sitting in the window ready for ice."

"So that's how you survive in the Arizona heat. Lead on."

"My car's here in the parking lot."

Mike nodded, and they made their way to the burnt orange Olds which Connie had bought with a loan from her grandfather. Behind the wheel she kept up the chatter, asking about friends at Reese and bringing Mike up to date on her brief term in the Air Force.

"You should plan your route this way more often, Mike. Jim drops in often on cross-countries. Tomorrow he'll be here long enough to broil steaks on my new grill."

Pulling up in front of the Bachelor Officers Quarters which had been her home since November, Connie cut the engine and turned to Mike. "I've been talking too much, but it's good to see you. I miss the gang at Reese." Mike was looking at her intently, but she had the strange feeling that he wasn't listening. "Mike, you're so quiet. Is anything wrong?"

"Just heat prostration, nurse, pure and simple," he said, but the quick shift of his eyes alerted Connie.

"Mike?" Connie watched the muscles in Mike's face tighten. Suddenly he opened the door and climbed out.

"Let's get that iced tea," he said as he came around the car, took her by the arm, and walked beside her down the sidewalk.

Connie's apartment was the first one on the south corner downstairs. It consisted of a small living room, dining area, kitchen, bedroom, and bath. She had felt at home here from the first day she turned the key in the door and stood surveying the government issue furniture. Having grown up in an Air Force family, living in government housing from California to Maine, she knew all the tricks of using colorful throw rugs, hanging pictures, adding J. C. Penney's bright slip covers to chairs and couch, and substituting new matching drapes and bedspread for Air Force issue.

Now, as she made her way to the refrigerator for ice, she was glad that she had bought the big planter of pink begonias and persuaded the lieutenant upstairs only yesterday to hang it in the front window.

Neither Connie nor Mike had spoken from the time they left the car. The soft slamming of the refrigerator door, the cracking of ice as it left the sides of the tray, the clinking of the squares in the two glasses, and the pouring of tea to the very top prolonged the silence between them. She pinched two leaves of mint from the pot sitting on the window sill and dropped them in the glasses. Then, picking up the cold glasses of tea, she walked slowly toward Mike. She handed him his tea, set hers on the coffee table, and settled herself on the sofa facing Mike.

"Mike?"

"Connie, Jim's had a bad accident. He's been air 'evacked' to Wilford Hall in San Antonio."

"Oh, no! Not Jim!"

The anxiety which had moved into Connie when she first knew something was wrong, now settled with a heavy thud deep inside her. She couldn't move, but her eyes never left Mike's as he talked on and on, explaining the details. She caught snatches of what he was saying.

". . .routine training flight. . .good student. . .high crosswinds. . .split second decision on touch and go. . .Jim didn't have time to make correction. . .student was killed. . .Jim's badly hurt, Connie. . .serious head injuries. . . ."

"It's not fair, Mike. He's only twenty-five."

"I know. He's a good pilot. But pilots live with danger, Connie. This could have been any one of us. It just happened to be Jim. It's the odds you take. Your dad's a pilot. He knows that."

Connie's dad had been an Air Force pilot for twenty-six years, but she couldn't remember being afraid for him. Flying was so much a part of her life that she must have blocked out the danger. Spot scenes flashed before her: she was playing in the waiting room of Base Operations at Offutt Air Force Base; she was clinging to a chain-link fence watching her dad climb out of a KC-97 at Malmstrom Air Force Base; she was sitting in the pilot's seat of a KC-135 on Air Force Day, the runway swarming with visitors.

The sound of Mike's voice saying he had to leave brought Connie back. Mike was standing, looking down at her. Connie stretched out her right hand, and Mike gave her a gentle tug to bring her to her feet.

"I'll run you down to the flight line," Connie said.

"No, thanks. A five-minute walk is just what I need before climbing into that T-38." At the door Mike patted her shoulder and then crossed the threshold.

"Mike, thanks. I. . . ."

"It's O.K., Connie," Mike said as he turned back. "Jim's tough."

Connie saw his eyes holding back the tears. "You're tough, too, Mike. Hang in there."

Connie stood in the doorway watching Mike as he walked rapidly down the sidewalk and turned left toward the flight line. The sound of the air conditioner working overtime reminded her that the half-ton unit was designed for cooling a space no larger than her apartment. Closing the door, she moved back to the sofa and let her body sink into the cushions.

This can't be happening, she thought. Not to Jim. We were just getting to know each other. It's just not true. She wanted to wipe Mike's whole visit from her mind. She didn't want to think about it.

But a terrible hurt was creeping in, and she was angry that she couldn't stop it. Helplessness, love, fear—all these merged with the anger and rolled into a hard knot in the pit of her stomach.

Mike's not a doctor, she thought. Jim's injuries can't be as serious as he thinks. He did say the accident was on takeoff. Jim hadn't even left the flight pattern.

Impulsively, she reached for the telephone on the end table by the sofa and asked the operator for Wilford Hall in San Antonio. Within seconds the hospital operator was contacting Jim's parents, who had arrived from Colorado Springs. They were in the waiting room. They will tell me Jim is going to be all right, she thought, as she tapped the phone with her finger and waited.

Connie recalled meeting the Alexanders just once. Jim's dad, an Air Force colonel, was efficient and distinguished looking. His pride extended to his son, who had chosen an Air Force career. Jim's mother was a quiet, gracious lady with short brown hair and green eyes. Her soft voice now came over the line as she said, "Hello, this is Bertie Alexander."

"Mrs. Alexander, this is Connie Ricks, a friend of Jim's. I don't want to bother you, but. . . ."

"Connie, Jim has spoken of you often. It's good of you to call."

"How's Jim?"

"His condition is critical. He doesn't know us. I. . .I really don't know much more."

"What do the doctors say?"

"They give us little hope, Connie."

"I. . .I just can't believe this. I. . .I. . .I'll call tomorrow."

Replacing the telephone, Connie buried her head in her hands. Dear God! I can't pray. He could die. Like my patient at Saint Mary's. That night in Lubbock, Texas, came crashing in on her. Working at her desk in the hall, she had looked up to see a bent old man trudging toward her. She rushed to Room 226 and found his wife, a tiny white-haired old lady, still and silent. Suddenly Connie had burst into hysterical tears and the kindly old man who had followed her back into the room put his arm around her shoulders to comfort her. "She's with our Lord, nurse. It's all right. Don't be afraid of death. Just a part of life, you know."

A part of life? Maybe for the old, Connie thought now, but not for the young. Not for Jim. He worked hard at the Academy. He deserves a full life—career, marriage, children. We both do. Death can't be a part of his life—not yet.

Hours later she went to bed. Living on an air base, she had learned to sleep with planes taking off and flying overhead. But tonight she couldn't sleep. Boysan and Tipper flights were practicing night landings in the T-38 jets, and she heard every maneuver. Jim had flown many such missions with his students. In the dark she reached for another Kleenex.

She couldn't picture Jim in a hospital bed. His 5′11″ frame was always full of energy—flying, having an after-work beer at the Club, taking a pretty girl out to dinner, teasing, laughing. He had once been serious about a girl named Phyllis, but his tour in Vietnam stretched the relationship to the breaking point, and a

Dear John letter had arrived shortly after his return to the States. He had told Connie about Phyllis in a rare serious moment one night over dinner. In fact, he had written "To Connie, who gave me back my smile" across an eight-by-ten hero shot.

Sleep came in the early morning hours, but it wasn't enough to refresh her before the clock on her bedside table released a jarring alarm at six. Raising herself on one elbow, she reached for the clock with her right hand and stopped the jangle. Her head fell back on the pillow, and the feeling of deep sorrow returned. Some nurse is checking Jim's temperature, blood pressure, and I-V tube, she thought. It was easy to picture the intensive care routine.

"Get moving," Connie said aloud to herself, forcing her body out of bed and into a cool shower. She dressed quickly in a fresh white uniform, put her hair up in a bun, and smoothed makeup on her face. After the finishing touches of a pale rust lipstick, she walked into the kitchen. The rosy-cheeked kid on the corn flakes box, wearing some kind of uniform, grinned at her through his wire-rimmed glasses. She didn't feel like smiling back, or eating corn flakes, or even drinking coffee.

Back in the bedroom, she perched her blue beret on top of her head and secured it to her bun on each side with long bobby pins. Reaching for her black shoulder-strap purse, she headed for the front door. The clock read six-forty.

Arriving at the hospital, she entered by the front door and started down the hallway. The boards creaked beneath her white rubber-soled shoes as she made her way past Ward 1 to Ward 2. It was only 7:00 A.M., but the thermometer some orderly had nailed to a post just outside a window of the walkway already registered 95 degrees. The sun had been drying the boards and baking the paint of the base hospital since it was first erected in 1941.

During the war cadet pilots by the dozens, taken off the production line because of stomach ulcers, influenza, valley fever, or an occasional air crash, had filled the beds in eight wings of the rambling one-story building.

Today, only four wards were needed to handle the minor problems of the diminished number of pilots being trained at the base and the backaches and menopause problems of the retirees and their dependents living in the area.

23 JOY IS DARKENED

The eight wards were tied together at one end by a single enclosed porch and were separated from each other by three-hundred square-foot patches of sandy yard space. Glancing out the window as she hurried to get from one air-conditioned ward to the other, Connie wondered why the Bermuda grass, which showed signs of agonizing struggle, didn't give up and let the sun and the sand have their way. Jim. . . .

Cutting her thoughts, Connie opened the door of Ward 2 and breathed in the cool air that bathed her body. She had never understood why the Air Force had abandoned the white cotton nurses' uniforms for polyester ones. Arizona was not polyester country.

"Good morning, Major Robbins."

"Good morning, Lieutenant Ricks. Let's get rounds made. The wing commander wants me at a meeting in an hour. Were you on duty when the young airman was brought in yesterday afternoon?"

"Yes, doctor. He injured his knee in a football game. Shall I get his chart for you?"

"No, I've read Captain Holcomb's notes. Looks as if we may have to send the patient over to Luke. If the X-rays show the cartilage is torn, he'll need surgery on that knee."

A knee injury. Connie wanted to laugh. What a blessing a knee injury would be, she thought. Jim's major injury is to his head. It is there we think, and know, and feel, and love. "Death is a part of life," the old man had said. What do I know about death or life?

Major Robbins was observing her with a puzzled expression on his face. He repeated something about rounds; Connie nodded, reached for her notebook, and followed him out the door of the office, past the linen room, past the bathrooms, and into Room 1.

"Good morning, Mrs. White. How are you feeling today?"

"Just fine, doctor. No complaints."

"Mrs. White, how about that headache you had yesterday afternoon?" Connie asked.

"Oh, yes—that," Mrs. White said, remembering.

"How long did it last?" Major Robbins asked.

"Oh, not over an hour, doctor. The nurse put a cold pack on my forehead and it felt better in no time at all."

"Very well. If it happens again, we'll get you some tablets for the pain. Have you been eating well?"

"Yesterday I was able to eat every meal."

"Good. Just keep that up and we'll have you out of here in a few days. At your age you have to be careful."

At your age, Connie thought. Age doesn't matter. Major Robbins nodded to Connie to get her attention. They went out, across the hall, and into Room 2.

"Good morning, Mary. How's my favorite appendicitis patient?"

"Not so hot, doctor. Did you really mean for me to be up walking yesterday?"

"That's right. Did you do it?"

"I did, but it wasn't easy. I've got to get back to my office, doctor. I'm a good secretary, you know."

"I don't doubt that, Mary, but it will be awhile before we return you to duty."

While they were talking, Connie was busy setting up the dressing tray. Major Robbins changed Mary's dressing and gave her instructions about walking. Then he and Connie went on past Rooms 3 and 4, which were unoccupied, and into the ward of twelve beds. A couple of back problems, a gall bladder case, two ulcers, and the young airman with the knee injury were waiting for the doctor's rounds before beginning their morning routine of baths, television, cards, and lunch.

Thirty minutes later Major Robbins was leaving Ward 2, heading for his meeting with the wing commander. Connie sat down at her desk to make the morning entries on patient charts before checking on the orderlies, who were already busy cleaning the hallways, restocking the linen room, and getting carts filled to begin changing the beds. She wrote the orders automatically, double-checking the entries she had made in her spiral notebook while making rounds with Major Robbins. Forcing her mind to focus on her work, she delayed any thoughts about Jim until the morning duties were completed.

Eleven o'clock came quickly. Knowing she had a half hour to herself, she relaxed her ban, and the impact of Jim's accident again flooded her mind. I know I can't eat, but I suppose I should try, she told herself as she headed for the hospital cafeteria.

The smell of broccoli and baked pork discouraged any attempt to consume a full meal. Walking to the drinks cart, she lifted a small carton of milk from the crushed ice, opened it, and poured the cold liquid into a frosted glass. Not bothering to sit down, she walked to a nearby window and gazed out as she enjoyed the coldness of the milk. It trickled down her throat, past the lump that would have stopped food.

The window framed the new red brick chapel just across the field from the hospital. As Connie stood staring at the announcement board to the left of the chapel entrance, she imagined she could make out the letters: "Sunday Services: 8:00 A.M., Protestant; 9:00 A.M., Catholic; 11:00 A.M., Protestant." Today is Friday. No service. No one will be inside, she thought. Her eyes drifted to the door, and she wondered if it would be locked. She turned, set her empty glass down on a table, walked through the swinging doors and found herself rushing down the hallway and outside. She didn't notice the heat as she ran across the stubby dead grass that was only barely visible in the sandy field. Reaching the steps of the base chapel, she slowed her pace and walked deliberately up to the door. When she pressed the thumb catch and pulled, the heavy brown door swung open and she went in.

Sun streamed through the windows on the left and fell on the mahogany pews, making the carpet and seat cushions seem a lighter blue than they really were. She went quietly down the aisle and into a pew near the front. It was cool here, quiet, and lonely.

God had always been in the background of Connie's life, like the Air Force. Her family was Presbyterian. When she was twelve years old, her name had been entered on the church register following confirmation class. She remembered that cold Sunday morning in Presque Isle, Maine. She had pressed her nose against the icy windowpane to watch the neighbors as one by one they had backed out of their snow-covered driveways and chugged down the street to the base chapel only three blocks away. Then she and her family had joined the caravan. She had thought a lot about God on that special Sunday. Kneeling at the altar she had felt the minister's hand on her head as he had prayed. She had walked back to her seat smiling. She belonged to Jesus. She was a Presbyterian.

During college days going to church hadn't seemed quite so important. She had tried the Catholic, the Episcopal, and the

Baptist. God was the same in all of them: most of the time far away. Since coming into the Air Force, she thought about God only on the Sundays she didn't have to work.

As Connie sat looking at the gold cross on the altar—plain, simple, empty—she wanted to talk to God and ask some questions. Why, God? Why? Why? She needed answers for Jim, for herself. But questions and answers from the Westminster Catechism intervened.

"What is the chief end of man?"

"Man's chief end is to glorify God and to enjoy him forever."

"What is God?"

"God is a Spirit, infinite, eternal, and unchangeable."

Infinite, eternal, unchangeable. The words rolled over and over in her mind. She was too numb to understand what they meant, but they soothed her, like a cool ointment on a deep cut.

Connie bent over and wiped the tears from her eyes with the hem of her skirt, rose to her feet, and retraced her steps up the aisle, out the door, and back to Ward 2.

TWO
CLOUDS DESCEND

The March wind, which ushered in the month of April, failed to bring rain. Connie tied a headscarf under her chin before getting out of her car for the short walk from the parking lot to the front door of the hospital. Her car was in Space 6, next to Major Robbins, and she made a mental note that he had not yet arrived. Pushing the lock button down on the door, she slammed it shut, dropped the keys into her purse, and started resolutely toward the hospital entrance. The wind was strong for 7:00 A.M., and the dust carried in mid-air swirled in the field next to Ward 2.

I wonder if Jim knows the sun is shining, Connie thought. Three weeks had passed since the accident. The almost daily call to Bertie Alexander to inquire about Jim had established a closeness between the two.

"There's no change in Jim's condition, Connie. I can't stand the waiting, day after day."

"I know, Bertie, but Jim's a fighter. You can't give up hope."

"Is there really a God, Connie? Do you believe there is?"

Yes, there's a God, Connie thought as she reached the door and pulled it open against a heavy gust of wind. But, Bertie, I don't know how to help you find him. I can't find him myself. She pushed her way inside, and the door slammed shut.

"Good morning, Lieutenant Ricks."

"Good morning, Colonel Marshall."

"You have two new patients on Ward 2 this morning. A ten-year-old girl was admitted for a tonsillectomy. Dr. Stone's doing it this morning. And Mrs. Green is back with that rash on her feet."

"I'll check on both patients right away."

"Lieutenant Ricks, don't you think you should do something about those short skirts of yours? It may be the style, but this is a hospital, and you are in the Air Force."

The young orderly working in the file directly behind Colonel Marshall stifled a snicker, and Connie could feel the blush coming to her face. "Yes, ma'am," she managed to say as she turned quickly and made her way toward Ward 2.

Tonight, she thought, I'll let the hems down. I don't want Colonel Marshall to turn down my request to go to Flight Nurse Training.

After the night nurse's briefing, Connie reviewed the new entries on patients' charts. Major Robbins still had not arrived when she put the last chart back on its rack, so she walked down the hall to Room 3. The door card read, "Betsy Tilman, tonsillectomy, Dr. Stone."

"Good morning, Betsy. I hear you're going to spend the day with me."

"Hi. Could I have some ice cream with my breakfast?"

Connie laughed, but before she could reply, Betsy's mother, who until now had not raised her head from the novel she was reading in the corner of the room, spoke out. "Betsy Tilman! I told you you'd have no food today." Betsy cried, Mrs. Tilman got angry, and Connie was trapped as a referee for five minutes.

Mrs. Green in Room 4 had her feet propped on pillows and issued a tirade of complaints about the night personnel. "No service at all," she said. "They completely forgot I was here." She really needs a medical team all her own, Connie thought when she finally escaped.

Noting that the cleaning cart had not been moved since she came on duty an hour ago, she went in search of the ward orderly, and found him just coming out of the supply room. "What's this, Airman Smith? Stretching a ten-minute cigarette break into an hour? Your cleaning cart is getting calluses from sitting so long, and you've got work to do."

Major Robbins was waiting in the office when Connie returned. Although late himself, he did not like being held up.

Connie promptly picked up her pad, apologized, and followed him across the hall to Room 1.

When Colonel Marshall called her in after lunch to discuss the request for Flight Nurse Training, Connie was more convinced than ever that she had made the right decision. She wanted her work to be important. Some flight nurses were even being sent to Vietnam. And she would be near Jim for five weeks. Wilford Hall was less than half an hour's drive from Brooks.

At four o'clock Connie was back in her quarters, dressing for a dinner at the home of the famous trio: Danny James, Jan Reeves, and Rich Engel. As she applied fresh makeup, she recalled how she had first heard of these T-38 instructor pilots.

Jim Alexander had flown in to Vandenberg Air Force Base, just six miles from Lompoc, California, the weekend she signed the papers that made her an official lieutenant in the Air Force Nurses Corps and ordered her to report to Williams Air Force Base. Jim scrawled the three names on a sheet of three-by-five notepaper and tucked it in the pocket of Connie's red sweater as he was leaving. "Look up these guys when you get to Willie, Connie. They're a unique trio, and you'll enjoy knowing them. Besides, you'll be glad to see some friendly faces in that desert."

She had met Danny first. Dozens of officers, young and old, married and single, gathered at the Officers Club bar after work. She had dropped in one Friday with Colonel Marshall and in a conversation asked if anyone knew Danny James. The next Friday she was perched on a bar stool when someone tapped her on the shoulder and said, "Hey, I hear you've been asking about me. I'm Danny James."

Connie spun around and faced a tall, handsome black man with an engaging grin on his face. Caught by surprise, she gulped, "Jim didn't—"

"Didn't tell you that I'm black?" Danny laughed. "It didn't cross his mind." Before the hour was over, Danny had introduced her to Jan Reeves.

Jan was one of those people who find it hard to be serious. His dark brown hair, cut in bangs on his forehead and hanging as long in the back as the Air Force would allow, made him look younger than his twenty-seven years. Five minutes later Jan had said, "Connie, the gang's having a party tomorrow night at our house. Want to come? I'll pick you up at six."

At the first party she had met Rich Engel, the third of the trio. Standing in the kitchen alone, snacking on fritos and bean dip, she had sensed eyes upon her and looked around to see a tall young man whose gray-streaked hair gave him a more mature appearance than his twenty-six years.

"Hi, I'm Rich Engel," he had announced as he walked toward her.

"I'm Connie Ricks. Have a frito."

Even now, remembering, Connie blushed as she recalled Rich's shattered composure when she shoved a frito stacked with bean dip into his mouth. He couldn't talk, and for some strange reason, she had experienced one of her few moments of shyness.

Avoiding large parties since Jim's accident, she had seen very little of Danny and Rich. But Jan called her often. Now, five o'clock retreat began just as he pulled up to her quarters in his yellow van.

Thirty minutes later the van stopped beside a mailbox set next to a narrow street and bearing the names of James, Reeves, and Engel beneath the number 2409. All the houses in this neighborhood on both sides of the street were walled in by eight feet of concrete and stucco, forming a straight line with the sidewalks. The flat roof of 2409 peeped over the top of the wall, and Connie heard the sounds of laughter and splashing of water from the pool that filled the front yard. As Jan swung open the big wrought-iron gate, Connie walked through, a bag of groceries in one arm and a loaf of bread, which didn't quite fit in the bag, in the other.

Her eyes took in the whole scene. A tall, slender blonde was bending over the edge of the diving board, both arms coming together at the wrists, fingertips touching, as she rolled into the transparent water. Coming up beside a long, blue rubber raft, she struggled playfully with the dark-haired lieutenant who claimed possession of it.

Across the pool Rich and another pilot were talking, sitting on the edges of their lawn chairs, drinks in their right hands. The conversation stopped for a minute, and Connie was conscious that Rich was looking at her. She was glad she had worn a skirt—a bright orange one with a sleeveless jacket which accented the white short-sleeved blouse. She knew the tan she had been working on for weeks gave her throat, arms, and legs just the right contrast to the orange and white.

"On to the kitchen, woman. There's work to be done," Jan teased, giving her a gentle push on the shoulder.

The house soon echoed with laughter, chatter, and an occasional shriek as the volunteer kitchen staff shucked, boiled, and buttered the corn, mixed a salad with everything from spinach greens to anchovies, wrapped potatoes in foil for baking, then delivered them from their prisons, cut them open, and stuffed them with cheese, chives, and condiments. Danny, Jan, and a vivacious dark-haired guy named Neal presided as head chefs around the grills. The blended aroma of barbecued crab legs and chunks of sirloin sandwiched between onions, green peppers, and cherry tomatoes drifted through the doors and windows.

The white tablecloth spread over the pool table was soon covered with appetizing food. At one end, a wicker basket held freshly baked rolls, and at the other end, two decanters of wine offered a choice for the hungry guests, who had now increased to thirty.

While everyone was going through the serving line, Connie drifted to the back of the house and into a room lined on two walls with bookshelves. In all the crowd, she was feeling alone. I shouldn't have come to the party, she thought. Somehow it doesn't seem right. Jim could have been here this weekend. But he isn't. Her eyes roamed across the books on the shelves, and came to rest on a Texas A&M annual, *The Longhorn.* She was holding it and reading Rich Engel's name inside the cover when she heard footsteps behind her.

"Connie? Checking up on my background?"

She turned toward Rich, who had changed from his swimsuit into a pair of tan slacks and a yellow knit shirt, then back toward the shelf to replace the book.

"I didn't know you were an Aggie," Connie said.

"I'm afraid I am. Hey, you're missing a good party."

Connie turned from the shelf slowly. "I'll be along in a minute," she said.

"O.K., but don't wait too long. We've got a bunch of hungry hounds here today, and the food is going fast."

As he left the room Connie wondered why she felt so shy around Rich Engel.

The sergeant in Personnel congratulated Connie as he handed her the orders, but she waited until she got back into her car to open them. Now the impact hit her as she read the words, forty copies deep:

1st Lt. Connie J. Ricks, 508-62-0595, to report to USAF School of Aerospace Medicine (ATC) Brooks AFB, Texas, to attend Flight Nurses Course 5OLY9761, starting 19 April 1972 and graduating 1 June 1972.

I'll see Jim, she thought. My schedule at Flight School will be tight, but I know I can relieve Bertie some. It must be hard, she thought, for Bertie to be there alone with Jim day after day.

Folding the orders, she put them back into the manila envelope and headed her car toward the gas station. While the gas was flowing into the tank, she pulled a map of the Southwest from the car pocket and spread it on top of the steering wheel.

"Hi there." Chuck Kavasack, a lieutenant she had dated once or twice, smiled at her as he leaned against her window.

"Hi, Chuck, where've you been?"

"Nowhere but Williams Air Force Base and surrounding area. But it looks like you're ready to travel. Where are you going?"

"I just picked up my orders for Flight Nurse Training School. I'm leaving for San Antonio early tomorrow morning."

"Congratulations, Connie. You'll love the silver wings. Meet me for lunch at the Officers Club and we'll celebrate with hamburgers."

"I really shouldn't take the time," Connie hedged.

"You can't turn down a free hamburger," Chuck argued. "Bring your map and I'll do a better trip plan for you than Triple A."

"O.K.," Connie said. "I'll meet you at noon." She waved to Chuck over her shoulder as she pulled out of the station. With stops at the hospital to say good-bye and at the Base Exchange to pick up hose and hair spray, she still had time to do some packing before meeting Chuck for lunch.

33 CLOUDS DESCEND

At twelve noon Connie paused in the door of the Officers Club Dining Room to survey the sea of uniforms seated four to a table and crowded in four to a booth. When Chuck stood up and waved from a booth in the far corner of the room, she started making her way through tables. Her hair, parted in the middle, swept behind both ears, and flowing straight down her back to her shoulder blades, swished only slightly from side to side as she eased past a chair, or turned to casually speak to someone she knew. She felt confident in the new apple green slack suit which she had bought at Goldwaters and was wearing today for the first time.

As she turned to her right, she saw Rich Engel looking directly into her eyes. Instantly Connie felt that same shy feeling which had come over her the night Rich had found her in his study.

"Connie Ricks. Well, how's the nursing profession?"

"The nursing profession is just fine." Rich had pushed back his chair and was standing with a napkin in one hand. Connie heard herself telling him about Flight Nurse Training and knew she was talking much too fast. "So, I'm leaving early in the morning. I'm meeting a friend for lunch who's going to mark this map for me."

Impulsively, Rich reached for the map she was holding, but drew back. "That's great, Connie. Have a good trip—drive carefully and all that. I'll tell Jan. . . ."

Rich sat back down as Connie made her way to the north corner booth where Chuck greeted her with a welcome smile. A few minutes later, Rich picked up his check and made his way to the cashier. Connie saw that he did not look back in her direction.

Two days of hard driving stretched between Williams Air Force Base, Arizona, and San Antonio, Texas. Keeping the car at a steady sixty miles per hour on Interstate 10, Connie let her thoughts center around Jim. Ed, her brother-in-law, was still safe, flying Chinooks in Vietnam. But Jim had not been safe in the Texas panhandle. She thought of her brother, Rick, just sixteen, torn between riding horses and flying airplanes. In two years the airplanes will win, she thought, and the Air Force, the danger, the war. . . .I wonder if Jim wanted to fly when he was Rick's age.

At five-fifty in the evening on the second day Connie pulled into the parking lot at Wilford Hall Hospital. The eight-hour drive from Van Horn behind her, she had stopped at Brooks Air Force Base only long enough to check into the BOQ room, take a quick shower, and change into a fresh navy blue and white chambray sundress. The dress was one of Jim's favorites. The square knot ties on the shoulders, trimmed with the same tiny red rickrack which ran around the scooped neck, gave her a youthful look to match the barefoot sandals. Driving around San Antonio on Loop 410 just as husbands were leaving their offices for home, secretaries were planning their evenings, and occasional impatient drivers were sounding their horns to keep traffic moving, Connie had taken just a half-hour to arrive at Wilford Hall. Now, getting out of the car, she locked the door and started walking toward the ten-story gray brick building.

The air was sultry. Even the clouds forming in the sky were not screening out the hot San Antonio sun. As Connie made her way through the parking lot at Wilford Hall, the asphalt was sticky under her sandals.

Suddenly a rush of anger checked her steps. God, he didn't have a chance. Do you know what it's like to go screaming down the runway in a supersonic jet out of control at 160 knots? God! It wasn't Jim's fault. So why? Why?

She had a strong impulse to run back to the car. Jim's voice as it had come over the phone four weeks ago was ringing in her ears. "I'll be landing at Williams at 1600 hours Friday, Connie. Can you pick me up at Base Ops?"

He's not at Wilford Hall, she thought. He can't be. He is. Oh, God, he never got to Williams. Hesitantly, she started again toward the hospital, weaving in and out of the long lines of parked cars. As she reached the cement walkway, the periwinkles lining the entrance nodded to her, the breeze tossing their delicate pink petals. She stepped over a small pile of grass which someone had pulled out of the flower bed and left to die on the sidewalk. Connie walked faster.

The big double doors of the hospital swung open just as she reached them, and a wheelchair pushed through. The welcoming grin of its young occupant caught her off guard for a brief second. Then she saw the unfilled pajama legs, folded up and fastened neatly with large silver pins. Both legs. Turning away quickly, she rushed through the door and into the lobby.

What's wrong with me? she asked herself. I'm a graduate nurse—an Air Force nurse. An Air Force nurse, she thought, who treats colds, stomach upsets, or an occasional broken arm from a hard game of Saturday afternoon intra-squadron football. At Williams Hospital patients don't die, or have serious surgery, or even run high temperatures.

Life had always been easy for her, a product of a loving family, good schools, an established church. Jim was an extension of the good things in her life, the one who for months had made the hop from Reese Air Force Base to Williams with striking regularity. It wasn't hard to trust God at Williams.

Wilford Hall was different. Patients from all over the States were flown here for special care: burn victims, accident victims, cancer victims—mutilated, suffering, diseased bodies. And Jim, a young Air Force jet pilot, was one of them.

Inside the hospital Connie worked her way past visitors, some with anxious faces, past bathrobed patients hobbling on crutches or pushing their wheelchairs, past the reception desk and the admitting office. Facing a wall of elevators, she pushed the up button and waited.

Twenty seconds ticked off her watch, then thirty. Two elevators were still going up. One stopped on the fourth floor. She decided to walk up. A few quick steps to the left of the elevators brought her to the door marked *Stairs* in black. Pushing it open, she faced a cavern of gray stone.

As she took the stairs one at a time, her right hand gripping the iron rail, her sandals clacked, echoing in her ears. Lights were on in the stairwell, but the hollow, vacant space increased her loneliness. She tried not to think as she climbed step after step, but fear began to knot the muscles in her stomach. Have I made the right decision? she asked herself. Should I have come? By the time she reached the landing on third floor, her knees were shaking hard. One more flight. She started up. For a brief moment the emptiness inside her cried out to God, but she reached the exit door. She didn't want to talk to God—not now, anyway. Pausing just long enough to steady her knees, she put her hand on the knob and gave it a half-turn. The door swung open.

Five minutes later she walked with Bertie Alexander toward Intensive Care. The nurse on duty, in a very professional tone, directed her remarks toward Connie. "Lieutenant Ricks, limit your visit to five minutes. Captain Alexander is in a coma." Then she added, "The head injuries are severe. He won't be able to talk. . . ."

"I know! I know!" Connie interrupted. Oh, God!

The stark whiteness of the room, the intravenous bottles on their six-foot stands, the groan of a patient in Bed 1, the smell of anesthesia, once an affirming part of her profession, now generated a chill that seeped through her skin as they approached Jim's bed.

What did I expect? she asked herself. Jim was in a Stryker frame, a padded metal basket used to turn unconscious patients over every hour. His head, encased in a turban of gauze, and the swollen face and eyes, veiled his identity. As she stood looking at his still body, she tried to separate the personal and the professional within her. Biting her lip as she blinked back the tears, she reached for Jim's hand lying limp beside his body. His cool fingers gave no response. His pulse was weak. She knew instantly that she couldn't reach him.

And she knew she could not erase the plane crash or cancel the split second of time that snuffed out one life and rendered another helpless. The cold reality was senseless, and destructive, and evil. She felt the anger fermenting deep inside her.

"God," she asked under her breath, "how could you stand by and not help Jim?" The silence only made her squeeze Jim's hand harder. No response. "Jim. . .Jim. . . ." Her lips moved, then closed. It was no use. The wall was hard, and solid, and high.

Bertie, whose four weeks of five-minute visits interspersed with hours of waiting had taught her more patience than she realized, gently took Connie's arm. Together they walked from the room.

THREE
NIGHT IS BLACK

Seated in Classroom D of the USAF School of Aerospace Medicine at Brooks Air Force Base with seventy nurses and flight technicians, Connie was struggling to keep her mind alert.

"Decompression sickness is any disorder caused by exposure to reduced barometric pressures," Major Gersbach was explaining. As the next slide flashed on the screen behind her, she lifted her pointer and continued. "Body cavities which normally contain gas and are affected by pressure changes outside the body are the middle ears, paranasal sinuses, respiratory tract, gastrointestinal tract, and teeth."

Jim's in surgery now, Connie thought. The call had come from Bertie early this morning. With pressure building up in his head, the decision had been made to operate.

"This could bring Jim out of the coma. Connie, please pray," Bertie had pleaded over the telephone. A faint glimmer of hope.

Connie glanced at her watch. Nine o'clock. Jim had been in surgery for two hours. If all went well, it could be several more hours before the operation was complete. But. . . .

Major Gersbach was down the list to the teeth. "Untreated cavities," she said, "can be the cause of dull pain at altitude, and the pain can persist after descent."

Connie looked at the exit door. It was ten or twelve feet from her, and she would have to climb over four people to get to the side aisle. I've got to get out of here, she thought, and get to a phone. Jim could die on that table. Bertie's alone. . . .

The class was standing up all around her. The lecture was over.

Rising quickly, Connie stood at attention while Major Gersbach left the room. With everyone moving toward the hall for the ten-minute break between lectures, Connie reached the pay phone just in time to see another nurse drop in her dime. The door of an office at the end of the hall stood open, and a phone was visible on the desk. Opening her purse to find the Wilford Hall number as she walked, Connie reached the phone, dialed, and waited. In seconds she heard Bertie's voice.

"Jim's still in surgery, Connie. I don't know anything more."

"Bertie, I'm through with classes at noon today. I'll be there by one o'clock."

The next two hours were an eternity. At ten minutes after eleven, Connie glanced at her notes. Only a few words covered what she had absorbed from Colonel Peterson's lecture: "Bends in and around bony joints of body. Oxygen at altitude won't stop pain. Descent will."

Colonel Peterson continued his discussion of neurological problems resulting from high altitude decompression. She tried to listen. "The brain may be affected by bubbles," he said, "but there is no particular pattern. The spinal cord, too, can be influenced. Watch for faulty vision, headache, partial paralysis, loss of speech."

Was Jim's surgery over? Connie envisioned herself walking into his room, over to his bed, looking down at his face. Jim slowly opened his eyes, then called her name. Bertie was crying. The scene faded and Connie was back to the lecture on altitude decompression.

An hour later she turned left out of the main gate onto Loop 410 and headed west to Wilford Hall. After two weeks, Connie had settled into a daily routine of class lectures, trips to Wilford Hall to be with Bertie, and hours of study late at night. Someone had to be with Bertie. She was alone now that Colonel Alexander had returned to Colorado Springs. But Connie wondered how long she could keep up the pace she had set for herself. When she arrived at the hospital, she found Bertie in the waiting room.

"Jim's back in ICU," Bertie said. "The doctor thinks he's holding up fine. He's sleeping now, but. . . ." Connie saw the hope in her eyes and sat down to wait with her. Neither of them said anything more.

A young sergeant in Army uniform sat in the blue armchair in the corner of the waiting room. He didn't even look up—he just kept staring at the floor and rubbing his hands, one inside the other. Two middle-aged women sat on the couch across the room and talked softly. Connie realized after a few minutes that she was listening to their conversation.

"Millie, I don't know what I'll do if John doesn't pull through."

"Don't think about it."

"I've got to. His heart won't take much more. I know that."

"Betty, you and John will be moving into that new house before you know it."

"We've been planning that retirement home for years. Do you know John and I have lived on twenty-two bases? He's had thirty-one years of service."

"Yes, I know. The Air Force moves us a lot faster than the Army."

A nurse walked in, and all eyes looked up. "Sergeant Miller, you can see your little boy now." The sergeant got to his feet quickly, knocking a magazine off the table with his elbow. He didn't take his eyes off the nurse as he followed her out of the room.

Bertie turned to Connie, shaking her head. "His little boy is only four and has third-degree burns all over his body. Med Airlift flew them in last night from Wichita, Kansas."

"How did it happen?"

"Car accident. His mother was killed instantly. The child was trapped in the car when it caught fire."

Connie got up, picked up the magazine the sergeant had knocked to the floor, and laid it on the table. When the nurse came in again, she turned to Bertie and Connie. "Captain Alexander is resting quietly now. If you'd like to come in for five minutes, you may."

As Bertie and Connie approached Jim's bed, they noticed his eyes were open. But they were staring straight at the ceiling, still, silent. He was breathing easily; the rising and falling of his chest was the only visible movement. Looking across the bed at Bertie, Connie sensed the hope in her eyes and watched as she gently touched Jim's face with her hand.

"Jim, Connie's here. You didn't know you'd wake up to see such a pretty girl, did you?"

"Bertie," Connie said softly, "he can't hear you."

"Jim, Dad's coming in tonight. He'll be here with you all weekend."

"Bertie. . . ."

"And when I talked to your brother last night, he had made an 'A' on his English exam. How about that?"

"Bertie. . .Bertie. . . ." She wasn't listening. Connie watched her face as she talked faster and faster. She knows, Connie thought. She knows Jim will never come out of the coma. He'll never recognize anyone. He'll never walk, talk, fly. The surgery only kept him alive. But is this living? What does being alive mean? Is death part of life? For a moment Connie thought she understood. Jim was enduring a living death. That was the truth.

Colonel Alexander arrived as Connie was leaving the hospital, and she heard herself saying, "I won't be back until Monday, Bertie. I have so much studying to do."

"I understand, Connie. You should get out some—see San Antonio. The river shops are interesting, and the Alamo."

The phone was ringing as Connie walked into her quarters an hour later. "Hi, Connie. I was beginning to think I'd been stood up. Can you be ready in thirty minutes?"

"Oh, Bill! I just got back from the hospital. To be truthful, I forgot."

"It's O.K. But a good rousing party is just what you need. Doctor's orders."

Perhaps this *was* what she needed. With just two drinks, she could forget about Jim. "All right, Bill, but I'll need that half hour."

Connie had met Dr. Bill Landry in the dining room at the School of Aerospace Medicine. A tall and lanky Texan, he had persuaded her to have dinner with him once before. After a quick shower, Connie slipped into a long red-plaid cotton skirt and a white peasant blouse. She recombed her hair, putting it back into a bun just off her neck. When Bill knocked on the door, she was ready.

"Well now, ma'am," Bill said, extending his right elbow, "let's get going to the Medics party. Then we'll sneak out and do ole San Antone till the wee hours."

Bill's plans didn't work out. Three hours later they retreated from the noise of the party to a booth in the Officers Club Dining Room. As dinner was no longer being served, Bill ordered two plates of bacon and eggs, but Connie left hers untouched. She felt compelled to talk about Jim. Bill ate his eggs in silence. The last bite gone, he laid his knife and fork across the plate and looked straight at Connie.

"I've listened to you for an hour, Connie. Do you think we're designed to live forever? The quicker you accept life as it is, the better off you'll be. 'Life's but a walking shadow, a poor player that struts and frets his hour upon the stage and then is heard no more. It is a tale told by an idiot, full of sound and fury, signifying nothing.' "

"I don't believe that."

"You better believe it, my friend, 'cause that's all it is. If you need convincing, spend a year in Nam. Or Wilford Hall is good for starters."

It was after midnight when Connie walked through the door of her quarters, but she dialed the California number anyway.

"Mom?"

"Connie, is anything wrong?"

"No, everything's fine. I just got home—Medics party at the Club."

"How's Jim?"

"He survived the surgery, nothing more."

"Connie, you've got to give him up to God. The Lord knows what he's doing. He sees the big picture that we can't see."

"I know. But right now I'm having a hard time. . . ."

"Believing that God is even there?"

"Something like that."

"Connie, Dad and I are praying for you. We love you very much."

"Good night, Mom."

Not only was Connie disturbed by questions about God, but recently she had found herself wondering if she had been right in requesting Flight Nurse Training. From the time she was a little girl she had wanted to be a nurse. Her plans were confirmed during her high school days when she worked after school for Dr. Wayne Marsh, a pediatrician in Omaha, Nebraska. She liked helping people.

But what she was learning now had little relation to Dr. Marsh's office. Every Military Airlift Command hospital plane carried a flight nurse, trained to care for all kinds of critically ill patients. From headquarters at Scott Air Force Base, Illinois, a network of planes operated around the world bringing patients to specialized hospitals. Emergency cases were picked up within an hour, those less urgent within seventy-two hours.

The first month of Flight Nurse Training had been packed with courses in medicine, surgery, orthopedics, psychiatry, experiences in the altitude chamber, and training in hyperbaric nursing. Next came survival training, which had ended with being ditched from a mock airplane into a huge swimming pool and then caring for patients on a raft for hours in the sultry San Antonio heat.

Today, Connie and her Aero-Medical team of three technicians were scheduled for a simulated training flight in the C-123 parked by Base Operations. On board, Connie checked the flight nurse's kit, medical equipment, oxygen, medications, records and charts against the patient manifest which she had received yesterday. She reviewed her patient placement chart before turning to Sergeant Kelly, Senior Medical Technician.

"Sergeant, is your checklist complete?"

"Yes, ma'am. All equipment is aboard and functioning properly. Baggage is aboard, cabin ready for patients."

"Thank you. Let's go." With a final glance at the I-V bottles racked in a neat row, Connie walked to the front of the plane. Eight names of people impersonating patients were on her manifest. Two ambulatory patients were boarded first, a woman diabetic and a man with eye injuries, patches over both eyes. After seeing that they were seated comfortably at the rear of the plane, Connie turned to the loading door directly across the aisle.

Within a short time six litter patients (all very good actors) were boarded: a burn case, a hypertension, a patient on a Stryker frame, another in a spica cast, and two duodenal ulcer patients. While Connie gave the patient briefing, other medical crew members were busy checking straps and seat belts. When Sergeant Kelly reported, "Cabin secure," Connie notified the pilot, and the plane began to move slowly down the taxi strip.

In the air over San Antonio Connie knew that every word she spoke to a patient, every order she gave to a technician, and every decision she made for the simulated en route emergencies were being carefully critiqued by her instructor seated at the front of the plane, her back to the flight compartment. But the flight was going well. Connie glanced at her watch. In ten minutes they would be landing.

"Nurse, nurse. . . ."

Connie moved quickly to the patient in the Stryker frame. "What seems to be the matter, Lieutenant Smith?"

"I'm hmmm. . . ."

Quickly she began to check the patient's vital signs as Sergeant Kelly came to her side. "Lieutenant Smith, Lieutenant Smith, can you hear me?"

"Yeah, but, hmmm. . . ."

"Sergeant, give him oxygen while I start an I-V."

Watching Lieutenant Smith sink into a coma—even a simulated one—brought Connie face-to-face with the reality of Jim's injury. She efficiently carried out her duties, but she couldn't stop the choking in her throat. As the old aircraft touched down on the runway, she was thankful that her instructor's attention was diverted by a minor emergency. The exit window at the rear of the old plane blew in. The panic created by the actor patients lasted only a few minutes, however, before the ground crew swung open the big exit door.

After Connie supervised the deplaning of patients, she stood outside watching Lieutenant Smith being loaded into a waiting ambulance. The air was hot and humid, and perspiration was running down her back. Suddenly she felt sick inside. I can't go to Vietnam, she thought, or live with suffering and uncertainty day after day after day. I can't.

A week later when orders came out, Connie was thankful that Colonel Marshall had requested her return to Williams. She wanted to go back. Rita Johnson, one of the nurses at Williams, had kept her posted on the progress of the new hospital. Building was on schedule, and the move out of the old World War II facility would begin late in September. Connie could almost smell the fresh paint and waxed polished floors of a new hospital.

And she missed her friends. "We're still gathering at the Officers Club on Friday nights for drinks and fun," Rita wrote, "and someone usually invites the gang over for Saturday night. The guys keep asking about you."

Now, as Connie put the last dress into her bag and closed the lid, she was glad she had her silver wings, but she did not envy the nurses who had flying assignments.

Driving into the parking lot at Wilford Hall for the last time, she wondered what decision would be made for Jim's care. The doctors had talked to the Alexanders about transferring him to a Veterans Hospital near their home. Walking up the cement sidewalk, she glanced once more at the multicolored periwinkles on either side. There was something about those thirsty flowers that pricked her conscience. Their heads were bowed in the hot noonday sun as if in submission to a higher power.

A short time later, sitting by Jim's bed, she tried to pray.

"God, I feel so dead inside, and numb. But I care so much." Reaching for Jim's hand, she slid her hand under his, interlacing their fingers. His hand was cool. She covered it with her right hand, spreading her fingers to warm his. "God, take care of Jim." Then, closing her eyes and struggling to keep back the tears, she whispered, "Please take care of me, too."

Suddenly she recalled the lifeless little sparrow she had held in her cupped hands when she was only ten. She had cried then, too, until her mother reminded her that God knew every time a bird fell to the ground.

"Even a tiny sparrow?"

"Yes," her mother had assured her, "he knows and he cares."

As tears escaped and ran down her face, Connie raised Jim's hand to touch her wet cheek. Slowly, she opened her eyes and whispered, "He knows and he cares, Jim." Returning the wet hand to his side, she knew her pain was becoming bearable, an ache with its edges dulled.

As she left the hospital the smell of a summer rain filtered through the dry, hot air. The first drops of water splattered the dust on the sidewalk, and the thirsty periwinkles lifted their heads.

The return trip to Williams Air Force Base took only two days. Connie drove straight to El Paso before stopping and cut the driving time on the second day by two hours. Driving through the gate at Williams, she smiled at the familiar face of Sergeant Taylor who gave her a salute and welcomed her back. Even the palm trees with their shaggy brown scales and the grass doing its best to survive without water looked beautiful as she guided the car straight to her quarters.

I wonder if Rita remembered to water my begonias, she thought as she pulled up in front of her door. At three in the afternoon all quarters looked deserted. Someone had turned her window air conditioner on, and the sound of the motor working hard reached her ears just as she opened the door of the car. She was glad she had let Rita know she was coming in today.

Gathering up her purse, a sack of cheese crackers and apples, and two empty coke bottles, she climbed out of the car. Her hands full, she gave the car door a shove with her elbow and started up the concrete walk.

As she opened the door of her apartment and stepped inside, she thought, Nothing's changed. It all looks the same—except my plants. They've grown. She made her way to the kitchen and got rid of the sack and bottles. When the phone rang she thought, That's Rita. But it wasn't.

"Connie? This is Bertie."

Connie's hand tightened on the receiver. Bertie wouldn't be calling now unless something was wrong. But Jim was fine two days ago. Fine? No, not fine. Bertie's not saying anything.

"Bertie? Tell me. What's wrong?"

The long pause made Connie wonder if Bertie was still on the line. Then she heard her voice, weak but calm. "Jim's gone, Connie. He didn't make it through the night last night."

Why didn't I call Bertie this morning? Connie thought. I've driven all day not knowing. She sat down slowly on the couch and took a deep breath to steady her voice.

"Bertie, Jim's all right," Connie said. She thought, How can I make her understand? "When I said good-bye to him, Bertie, I knew. God kept Jim here long enough to settle a few things with him—that's all. Bertie, he's all right. He's in the hands of a loving God. I know."

FOUR
IGNORE THE CLOUDS

Chuck Kavasack stood in the door of Connie's quarters at Williams Air Force Base, hands on his hips, smiling as he watched her struggle to close an obviously overfilled cardboard box with two-inch masking tape. The tape popped open each time she removed the pressure of her hands. After enjoying her annoyance for a moment, Chuck volunteered, "Why not forget it, Connie? The box can't budge when it's crammed in the car, and an open box of towels is no real threat anyway."

Connie's need to be tidy gave way to her practical nature, and she laughed. "You're right. I just hate to see a packing box defeat me."

Pushing the hair out of her face, she rose to her feet and surveyed the almost empty room. Bedroom, bath, and kitchenette had been stripped hours ago. Now the last few boxes were sitting in the middle of the living room ready to be carried to the car. The bare, scuffed linoleum would need some wax before a new occupant could move in. The hum of the refrigerator reminded her to take the few steps into the kitchen and pull the cord. Opening the door, she took out two canned Cokes. Then, leaving the door ajar, she stood looking out the window. No one was playing baseball today in this July heat. The ground was parched, and the sun glared on the Base Exchange warehouse across the field.

Chuck had just put the last box in the car and returned when Connie walked back into the living room. "Here's your reward for helping me all day," she said as she handed him a cold Coke.

"I'll take it, but not in full payment. You're having dinner with me tonight at the Salt Cellar."

As they stood in the empty room drinking their Cokes, Chuck said, "Connie, why are you moving off base? The hospital is four blocks away, the Club three, and you're letting yourself in for an hour of driving every day. When you're on night shift—"

"Fine time to debate the pros and cons of this move, Chuck. Do you want to move me back in?"

They both laughed. How could she explain? Would he understand that she needed some time to herself, away from the base, the hospital, the Club, the young flying officers who were her friends? Flying to Colorado Springs a month ago for Jim's funeral, she had begun to experience the certainty of death—and the importance of finding some purpose for her life. As she had sat in the Air Force Academy Chapel and allowed her eyes to imprint on her mind the uniformed flyer on the stained glass window, she had felt the chilling permanence of the separation between life and death. How could death be a part of life, as the old man had said? It was the end of life. She had wanted to comfort Bertie the day they stood by Jim's grave, but she had had nothing to say, nothing to give.

The month of August was unusually hot, even for Arizona. The thermometer climbed to 115 degrees most days, and evaporator coolers strained to push the mercury down inside the older buildings on base. The new hospital was nearing completion, and Connie looked forward to being a part of the move in early fall. Her off-base apartment had not provided the quieter pace she had anticipated. Days were an endless succession of eight-hour shifts at the hospital, cocktails and dinner in Phoenix with Jan, Chuck, Bill, or some other young flyer, home, an hour of television, and sleep.

On Friday night in the bar at the Club, Connie was a little surprised to hear Colonel Maglione say, "Find yourself a G-suit, Connie, and I'll take you up in a T-38 on Monday. Takeoff time is 1100 hours. We'll see how tough you are."

"All right, Maggie. That's my day off." Connie could feel the eyes of half a dozen pilots on her. She wasn't afraid to fly. "I can make it," she answered confidently. "I know where I can borrow a G-suit."

No one spoke for a moment. Then Chuck lifted his glass. "To Connie, and the flying fraternity!"

Later that night as she lay in bed listening to the sound of planes overhead, she remembered her one flight in a T-38 last January. She had suggested to a group of instructor pilots that they take her along in the fly-by salute to graduating Class 70-05.

"A nurse in a fly-by? I don't know, Connie. You'd probably be scared stiff."

"Forget it, beautiful. We'd love to have you and all that, but you better stick to your nursing."

"We'll meet you afterwards for lunch and tell you all about it."

Major Venturine had been different. "Sure, Connie. Why not? You're an Air Force nurse. I think we can get you a flight. But you'll have to find a flight suit." No problem there. She had wangled a green flight suit from the Individual Supply Store sergeant, with CONNIE neatly printed above the left upper pocket.

Graduation exercises were scheduled for 7:30 A.M., but she had arrived at the squadron an hour early and walked boldly into the Boysan briefing room. Her confidence had evaporated, however, when she met not the friendly greeting of her social companions in the flying fraternity but the cold formality of a group of professionals already in the process of being briefed on the morning's flight by Major Venturine, the flight leader. Taking a seat next to the wall, she had felt conspicuous, being the only woman in the room.

"You all know the formation," Major Venturine had said. "Major Ferguson and Captain Engel will lead. Captain Craig, you and Captain Brown are two. Captains Anderson and Smith, three, and Captains Flair and Thompson, four. I will lead the second element. Lieutenant Ricks will be in the back seat."

Major Venturine's voice had been the focal point in the room, and Connie had tried to catch every word. "We have two extra ships over here to fly chase. They'll fill in for you if you have any problems. We'll start after a check-in at 8:05. Let's do a Superstition Two departure, and go into holding pattern over Superstition until we get the timing down right. I've already coordinated with Phoenix Approach Control."

Connie had begun to relax a little as the briefing went on. "We'll have radio contact with someone on the ground so we can adjust our speed and hit the field just as the parade marches by the grandstand. The Tweets will be in front of us till we get to the edge of the field, and then we'll come down in front of them."

Tweet pilots were second-class citizens to any self-respecting T-38 driver. Having friends in both groups, Connie enjoyed the rivalry. Major Venturine concluded the briefing. "We've done this before. Hang in tight and make it look good."

Walking with Connie to their plane, Major Venturine had explained last-minute details. Smiling, the crew chief checked her seat belt and helped her adjust her chest straps and oxygen cord. In the air, Connie had enjoyed the excitement of the radio calls and the sound of her own voice when Major Venturine allowed her to respond to the tower once or twice. When the plane turned slightly, she pressed her feet against the floor and held tightly to the seat to relieve the feeling that she was falling through the clear glass of the cockpit.

The Tweets were on tower frequency, flying ten miles or so in front of the T-38s. Suddenly she had heard Major Ferguson direct Boysan Flight to channel one. "Boysan Flight check-in. Red lead."

"Two."

"Three."

"Four."

"White lead."

"Two."

"Three."

"Four."

"Blue lead."

"Two."

"Three."

"Four."

"Yellow lead."

"Two."

"Boysan Flight with fourteen of the free world's finest on final for fly-by."

The time had come to swoop over the field. Moving in fast,

the Boysan Flight had caught the Tweets in the jet wash and bounced them soundly.

Now, Connie opened her eyes in the dark and smiled as she remembered the half-angry joshing thrown back and forth between the pilots over lunch after the morning's festivities. And she had accepted their congratulations. "I'll call about that G-suit in the morning," she said to herself as she dropped off to sleep.

Connie dialed Jan Reeves' number at 10:00 A.M. and heard the phone ring twice before Rich Engel's voice said, "Hello."

"Is Jan there?"

"No."

"Is Danny there?"

"No."

"Do you have a G-suit I can borrow for a T-38 flight with Colonel Maglione on Monday?"

"Maggie is going to take you up in a T-38?"

"That's right."

"You can borrow my G-suit. What's your takeoff time?"

"1100 hours. When can I pick up the suit?"

"It's in my locker in Boysan. I'll be teaching at 1100. Ask any of the IPs in the flight. They'll get it for you."

"O.K. Thanks, Rich. See you."

"You're welcome. Have a good flight."

As Connie hung up the phone, she thought, I wonder if he likes lasagna.

Sunday morning Connie went to the eleven o'clock service at Grace Community Church for the third time since moving off base. When Tom Watson dropped by her apartment in the afternoon, the pastor's sermon was still on her mind.

"Are you interested in religion, Tom?"

"Oh, I've thought about it. I go to church now and then."

"Do you think God has a plan for every person's life?"

"I really don't know him well enough to ask, Connie."

"I think I'm afraid to ask because I'd want it to include the Air Force, my nursing, marriage, children. God might have other plans."

Tom laughed and picked up a book lying on the table: *Know Why You Believe* by Paul Little. "You're really serious about this, aren't you?"

"Yes, I guess I am."

Rich was not in his classroom when Connie peered in the door. The shuffling of students' feet in the hall as they left the building for lunch created a muffled background for their laughter and question-and-answer decisions about lunch, dates, study, movies. Connie walked down the hall to Rich's office, through the door, and over to his desk, stacked neatly with flight manuals and student papers. Making herself comfortable in his swivel desk chair, she reached for a yellow slip of paper from a square plastic holder and a pencil from its beer can container.

Dear Rich,
Wow, what an airplane! The G-suit worked perfectly and needed only minor adjustments. Since you're batching it this week with your roommates gone, how about dinner at my place Saturday night?

Connie

At 5:05 that afternoon the phone rang in Connie's apartment.

"Connie? Rich Engel. Dinner Saturday night will be fine."

"I'm glad you can come. I put your G-suit back in your locker."

"Thanks. You enjoyed your flight?"

"It was neat."

"Well, see you Saturday."

The week was a busy one for Connie, but when Rich knocked on her door that Saturday at precisely 7:00 P.M., she was ready. The crisp green salad was chilling in the refrigerator, cheese was melting on the lasagna in the oven, and the sourdough bread was already in the bread warmer, cut and buttered. Picking up the glass of scotch and water, she turned to give the table a final glance: white tablecloth, white candles in the crystal candleholders borrowed from her neighbor, two plates of dark green stoneware, two green waters and wines to match, knives, forks, spoons, and gold-fringed napkins. The yellow and gold carnations, arranged in a small bowl with baby's breath, added just the right touch of color.

"Not bad," Connie congratulated herself. As she moved toward the door, she hoped belatedly that Rich liked lasagna. Her hand on the knob, she took a deep breath and opened it.

"Hi, come on in."

"Thank you. You're looking very pretty tonight. I like your dress." For the occasion Connie had bought a long brightly flowered cotton dress with a sash which looped on the side and hung almost to the hem.

Connie led the way to the comfortable gold wing chair at the end of the coffee table and motioned Rich to sit down. She seated herself on the sofa.

"Oh, I'm sorry. This is your drink!" she exclaimed, thrusting the glass into Rich's hand.

"Thanks. How about yours?"

"I guess I forgot," Connie laughed, managing a self-conscious smile as she excused herself and walked hastily to the kitchen.

The rest of the evening clicked according to Connie's well-planned schedule. The fall *Journal of Nursing*, the latest *U. S. News*, and Wouk's *Winds of War*, which she had placed on the coffee table to impress Rich, provided enough conversational topics to span the time between the ice cream dessert and the clock's stroke of midnight.

Before Rich left, while he talked about Kissinger's policy of foreign aid to third world nations, Connie wondered if he would kiss her goodnight. But he didn't.

"The dinner was great, and the company even better, Connie. Thanks for asking me. You can borrow my G-suit anytime."

"I'll remember that. The next time I get invited up in a T-38, I'll give you a ring. Goodnight."

On Tuesday afternoon after work Connie ran into Danny in the Base Exchange. "Hey, Connie, what's going on between you and Rich?"

"What do you mean? He came to dinner Saturday night, that's all."

"Don't give me that. I know Rich Engel. Last night he sat down to read *U. S. News*, and an hour later he was still on page one. Now a guy who has three credits from Arizona State in speed reading should do better than that. He finally asked me if I thought Jan would mind if he took you out to dinner."

"He called me. We're going out Friday."

"Rich is a great guy."

Connie just smiled.

On Friday night, Connie sat looking at Rich across the table in a corner of the Fig Tree restaurant in Scottsdale. The candle-light was flicking shadows on his face, and his ice cream was melting. He was talking about the Avenger, his turbo-charged trick car which he had paid a friend to build for him while he was in Southeast Asia. Polished and shining, the car which looked like a Corvette had been the major topic of conversation since Rich picked her up for their date.

"The frame is all handmade to the basic dimensions of a Volkswagen, but the suspension front and rear are Corvair. Of course the visibility isn't very good. I suppose Ralph Nader would go into shock if he ever rode in it."

When the candle burned down to an inch above the holder, and their coffee cups were empty for the fourth time, the restaurant was closing. "Rich, I'm leaving for a four-week obstetrics and gynecology course at Wichita Falls, Texas, in ten days," Connie said. "I'll come back to our new hospital as a GYN clinician."

"That area of medicine interests you?"

"At the moment, yes. Caring for women who are expecting babies will be a nice change for me. We need a nurse trained in that area, and Colonel Marshall thinks I'm the one."

"Are you driving to Texas?"

"Yes, I am. Want to come?"

"You're joking, but seriously, I'm flying to Connecticut for Neal and Cheryl's wedding that weekend. It's possible that I could drive you to Wichita Falls and fly from there if that would be of any help to you."

By the time Rich eased the Avenger to the curb in front of Connie's apartment, their plans were all made. For some reason Connie felt that a whole new chapter in her life was beginning. The cool desert air blew through the open windows of the car, bringing the smell of sage and honeysuckle. She wondered if Rich would kiss her goodnight. He did. It was September 8, 1972.

FIVE
LOOK TO THE CLOUDS

On Sunday afternoon Shepherd Air Force Base at Wichita Falls, Texas, looked deserted. Drizzling rain was freezing on the tree limbs, and Connie's lightweight coat was not warm enough for this unexpected weather, even with the car heater going full blast. She drove around several blocks after dropping a letter to Rich in the Post Office, but not seeing a familiar face, returned to the BOQ room. The phone was ringing as she opened the door, and Rich's voice came over the line.

"Hi! How's my favorite OB/GYN nurse?"

"I know. Your only OB/GYN nurse. Hi, Richard."

"Oh, yeah. My thinking is probably fuzzy from the dazzling sunshine reflecting off the pool. I've been out here toasting for the last couple of hours."

"You turkey. It's sleeting here in Wichita Falls."

"That's the breaks. Say, have you been following McGovern's comments the last few days?"

Thirty minutes later, after a thorough review of the presidential campaign, Connie hung up the phone. She resolved to keep current on U. S. politics.

Monday morning Connie shifted into the routine of the OB/GYN clinic. Dr. Edward Yosowitz's efficient manner intrigued her, and she could see the necessity of moving patients quickly from waiting room to consulting room, to examining room, back to consulting room, and out of the clinic. It was now 11:00 A.M., and ten more women waited to be seen.

One, a mousy little woman, sat in the corner of the room holding her hands in her lap, not moving at all. "Mrs. Smith, the doctor will see you now." Connie smiled, but the meek little woman only turned slightly when she heard her name called, rose to her feet, and moved with reluctant steps into Dr. Yosowitz's office. Closing the door quietly, Connie motioned Mrs. Smith to a chair and waited for the doctor to speak.

Several minutes passed as Dr. Yosowitz flipped the pages of Mrs. Smith's medical record, scanning each page and nodding his head at intervals. With his hand on the last page, he looked up and began in a firm voice.

"Mrs. Smith, you have refused surgery although you have a tumor the size of a baseball growing in your uterus."

"Doctor, I have three children—all so young. They're just in grade school. What if I died? There's no one to take care of them. Can't you give me some medicine? I feel fine."

"Mrs. Smith," Dr. Yosowitz said, his voice growing stronger with each word, "two of our best doctors have explained your condition to you. No medicine will take away that tumor. It's growing larger every day. It won't stop growing. If you don't have surgery soon, you'll get much worse. Then who will take care of your children? Mrs. Smith, you must face the truth."

Dr. Yosowitz's voice filled the room. The woman was sobbing and Connie's face was drawn in concern. Dr. Yosowitz's eyes met hers. She took a Kleenex from the box on his desk and handed it to Mrs. Smith, at the same time placing an arm of support around her shoulders.

"All right, doctor, all right," Mrs. Smith said. "How soon?"

"Let's do it right away, Mrs. Smith. Lieutenant Ricks, see when we can schedule Mrs. Smith for surgery. Don't make it later than Friday."

"Yes, sir."

"Lieutenant Ricks will tell you when to check into the hospital, and she will answer any questions you may have," Dr. Yosowitz now explained to the trembling woman. Then he added in a kind but still firm voice, "You'll be fine, Mrs. Smith, just fine."

But Mrs. Smith wasn't fine. Assisting with her surgery on Friday morning, Connie froze as she heard Dr. Yosowitz exclaim, "Oh, no! Look at this. She's covered with it. We're too late." His brow was wrinkled above his surgical mask.

Connie bit her lip. It was quivering. How did I get so involved? she asked herself. I didn't mean to. "How long does she have, doctor?"

"Maybe a month—less if she goes home." Connie was sure that she saw tears in his eyes. Looking up at Connie, he said, "I'll finish here. You can go. Meet me in my office at three."

The next week Dr. Yosowitz drove his new protégé hard. "What would you do, Connie, if a twelve-year-old girl came to you asking for an abortion?"

"The law states. . . ."

"The law. Come on, Connie. Sure you know the law. But how do you feel down deep inside about abortion?"

"I don't know. I honestly don't know."

"Then find out. Think it through. THINK. Let's take another case. That mother today who brought her fifteen-year-old daughter, Angie, for the pill. I gave her a prescription. What would you have done, Connie?"

"I. . .I. . .I don't think. . . ."

"Do you want her pregnant? In a few months she'd be back here in OB clinic. How do I know that? I know people like Angie—read the statistics. You can't stop them—just try to keep them from bringing unwanted babies into the world."

Between patients, Dr. Yosowitz brought up case after case.

"Connie, let's talk about Theresa Brown, the woman whose baby is due in two months. She has endometriosis. Predictable complications have cut the mother's chances of survival. We'll do a cesarean. How would you have handled the case a year ago when Mrs. Brown came for help in getting pregnant?"

"It would have been a difficult decision to let her risk her life."

"That's right, but it's her life, her choice. She wants that baby. She's a woman. She feels her purpose is to bring life into the world. She has a right to give life, Connie, and I have the skill to help her."

Dr. Yosowitz's chance to help her came sooner than he expected. The call came from Colonel Brown at 3:00 A.M. Mrs. Brown was in labor. At 11:00 A.M., Connie stood watching the skillful hands of Dr. Yosowitz fight for the lives of Theresa Brown and the infant within her who insisted on entering the world ahead of schedule.

"Scalpel."

"Scalpel."

"Clamp."

"Clamp."

"Suction."

"Suction."

The sound of the instruments slapping into the doctor's hand echoed in Connie's ears.

"Theresa," the doctor was saying, "I'm making an incision in your tummy. I'm clamping it off little by little to get down to the uterus where your baby's waiting. The incision's so little it won't even show with your bikini."

Theresa smiled. A sheet hung over a metal bar dividing her face from the rest of her body. Connie stood across from Dr. Yosowitz, observing scenes on both sides of the curtain. The surgical team knows their business, she thought, and Theresa is doing fine, so why do I feel so anxious?

"Theresa, we're down to the uterus. We'll have to wait and time this right—between contractions."

The silence was deafening.

"O.K. Let's go in. Clamp—suction." Every hand went into action.

"Theresa, you O.K.?" Connie heard herself ask as she wiped beads of perspiration from the patient's forehead. The spinal anesthesia had made Theresa numb to physical pain but not to mental anxiety. Connie concentrated her attention on Theresa whose eyes were unable to pierce the white sheet. A mumbling of voices now came from within the closed circle of professionals. Connie sensed Theresa's mounting fear.

"Relax, Theresa, your baby's almost here," she said.

"Nurse, my mouth's so dry."

"Does that help?" Connie asked as she touched a cool damp cloth to Theresa's lips.

"It's a girl, Theresa, a fine baby girl!" A baby's cry came from behind the curtain with Dr. Yosowitz's announcement. Connie glanced up to see a little, wrinkled crying baby, still coated with a protective red and white covering, and still attached to the long umbilical cord.

Theresa's face was wet with tears, and radiant. Her eyes on Dr. Yosowitz holding her baby daughter, she said, "Doctor, I just don't have the words to thank you. Oh, I could just kiss you!"

Together Theresa and Connie watched the pediatrician, Dr. Nevin, clean the baby and wrap her in a blanket under a warm light. He took the baby's footprints and blood sample, and put drops in her eyes. Handing Connie the identification bands, Dr. Nevin smiled at her as she placed them around mother's and daughter's arms.

"Mrs. Brown," Dr. Nevin said, "Lieutenant Ricks and I are going to take your baby to the nursery now and let you and Dr. Yosowitz finish up here."

Connie saw Theresa's eyes flood once more with happy tears as she said, "Thank you."

Dr. Nevin lifted the baby from the warm table and placed her in Connie's arms. Connie had observed many births, assisted in a few, but somehow this child was special. As she held the little girl and turned to place her in the incubator, she could feel the warmth of the tiny body against hers. The wrinkled face looked more like a dried-up apple than a baby's face, but it was beautiful. "God made you, little one," she whispered to the child. "You're going to be loved and cherished, and you're going to live. Do you hear me? You're going to live."

Just before Thanksgiving, Dr. Yosowitz relaxed his rigid training long enough to let Connie know she was going to make it. The last patient had left the office when Dr. Yosowitz reached for his pipe and motioned her to a chair across from his desk.

"You're good, Connie," he began, "very good. Another month and I'll have you shaped into the best OB/GYN nurse anywhere around."

Connie was silent. She knew she had learned more than skill from Dr. Yosowitz. He had taken her eyes off of death, and forced her to focus on life.

"My tour of duty with the Air Force will be up next summer, Connie," he said. "I'm going to Houston to set up a practice with another doctor." He tapped his pipe, struck a match, and lit the tobacco. "How would you like to come with me, Connie, as my right-hand nurse? I can promise you a good salary—a lot more than you're making in the Air Force."

"I didn't know a nurse could ever make much money," Connie laughed.

"Stay in the Air Force and you never will. You would scrub

with me in the mornings and see patients with me in the afternoons. Give this offer some thought, Connie. I need a good nurse." He puffed on his pipe and stood up. "Incidentally, Connie, Brenda asked me to invite you for dinner Friday night. I'd like you to meet my wife and my beautiful little girls."

Because of her tight training schedule, Connie had not visited her grandparents in Cameron, a five-hour drive away, since coming to Texas in September. Now on Highway 281 headed south, she knew that by avoiding Fort Worth traffic she could arrive in Cameron in time for Thanksgiving dinner. She thought of Grandma and Granddad Wilson, married fifty-four years this past September. They had reared three girls with the help of the First Baptist Church congregation, and moved to their small ranch just north of town twenty years ago.

Connie was born in Cameron Hospital, and the summer and Christmas visits to the ranch over the years gradually gave her a feeling of coming home. It was here that she had come to know her aunts, uncles, and cousins. She remembered how fresh the hay smelled when she had hidden in the barn loft waiting to be found. Once her cousin Tommy, only five, had fallen through a hole, and fifty bales of hay had to be moved to rescue him. Granddad stocked the pond with fish when Cindy, the first grandchild, was old enough to hold a pole, and Connie had learned early how to persuade her cousins Jay and Wayne to bait her hooks. A sandbar on the river, called Buzzard's Roost, was a favorite place to swim. As each of the nine grandchildren graduated to swimming age, Grandma packed sack lunches of fruit, jam sandwiches, and Pepsis for trips to the river. The Pepsis were kept cool in the river water, but the sandwiches usually disappeared long before lunch time.

Connie remembered her first horseback ride alone on Charlie, who mistook her moving of the reins for his cue to round up a stray cow and calf at the far end of the pasture. And she recalled her first time on the tractor when Granddad had climbed down and left her in charge of the big machine.

During college years at the University of Texas, weekends at the ranch had helped to ease the loneliness of being away from her parents in California. Connie had made many important decisions while talking with Granddad late at night on the front porch. By then, the fireflies, once in danger of being captured and put in jars, flitted about freely in the yard, and Connie had known she was growing up.

At eleven-thirty Connie turned off Interstate 77 at the Cameron airport and heard the crunch of the gravel under her tires. Crossing the half mile over to the Salem road, she turned left and headed for the ranch.

Everything looks the same, she thought, as she approached Granddad's property line. The water tank on the hill had "Wilson Ranch" painted on it in huge black letters. Water for the house came from an underground spring and was often scarce during dry seasons. White-faced Herefords grazed on tall green grass which stretched past the pond lined with spreading oak trees. As the house, sitting back from the road, came into view, Connie could see the Adams' and Beavers' cars in the driveway and Granddad, and Lanier and Jim, her uncles, walking toward the barn. Mollie, the brown and white sheep dog, announced her arrival with loud barks even before she pulled into the yard and stopped.

The whole family came out to greet her: Jan, Donna, Wayne, and Tommy Adams, Beth and Jay Beavers. Her aunts, Rita Adams and Gwyn Beavers, waited on the porch with Grandmother. Connie hugged each one.

"Jan, how's Texas U? Donna, I hear you're going into physical therapy. Tommy, how'd you get so big?"

"For a sixth grader, Tommy's doing all right, Connie," Beth said. "He's got two girlfriends." The youngest of the cousins, Tommy took a lot of teasing.

Reaching the porch, Connie looked at Grandmother, smiling and waiting patiently. "Connie? Well, I declare." Unable to see more than outlines since surgery for a brain tumor more than thirty years ago, her eyes strained to see her granddaughter.

"Grandma, you've lost ten pounds and you're as beautiful as ever," Connie laughed as she went into her arms. The smell of Cashmere Bouquet bath powder, which Connie always associated with Grandma, was still there.

"Connie, I'm so glad you're here."

Connie noted the silver gray wavy hair which framed her

grandmother's soft face. There were no black streaks in the gray now.

The house smelled of turkey, dressing, and pumpkin pies. Later the family sat around the dining table, all heads bowed, as Granddad asked the blessing. Sensing a special warmth in his voice, Connie opened her eyes just as he wiped a tear from each side of his face. He looked much younger than his seventy-four years, a tower of strength to the whole family, but at times like this Connie knew his soft heart.

"Lord, you've really poured out your blessings on us this year." Connie saw his hand reach for Grandma's as he went on to thank the Lord for watching over his family. "I have 'no greater joy than to hear that my children walk in truth,' " he concluded.

Walk in truth, Connie thought. Am I walking in truth? Dr. Yosowitz's offer had been hammering at her conscience. She had pictured herself in a nice apartment in Houston, a closet full of designer fashions, money in the bank. She liked Dr. Yosowitz. He demanded excellence and working for him would be a challenge.

But Rich's phone calls were a frequent reminder that she was falling in love with an Air Force pilot who could offer her more love than money or security. In just five years with Dr. Yosowitz, she thought, I could save. . .travel. . .and I'd love the work.

That night, after the Adams had returned to Houston, the Beavers had left for Frisco, and Grandma had gone to bed, Connie sat with her granddad on the front porch. The moon gave the yard a soft glow, and the fireflies flicked their lights off and on at regular intervals. A bullfrog from the pond joined the crickets in their evensong.

"Granddad, what does it mean to 'walk in truth'?"

The long silence that followed made Connie wonder if he had heard the question. Finally he raised his head and looked up to the star-speckled sky, surveying the whole horizon before he turned to her.

"Well," he began, "we can see a good example in your great-grandmother Bollinger. You never knew her, Connie, because she died before you were born. She never got outside Texas—was born just a few miles from here over in Marlow, and she's buried in the Marlow cemetery. She never owned a home, and came to live with us when your great-grandpa died. She helped raise your mother, Rita and Gwyn, especially after your grandma's surgery. She was 'Aunt Nettie' to everybody, nursing the sick, making quilts and sewing and cooking for anyone who needed her. She read her Bible every day and had a talk with the Lord to find out what he wanted her to do. Then she just did it. That's 'walking in truth,' humbly and unselfishly serving others the way the Lord has mapped out for each of us.

"You're a lot like your great-grandma, Connie. She never had a course in nursing as you have, but she had real compassion for people. I know you have too. Don't get sidetracked and start looking for honors, promotions, material things for yourself. You just find out what work the Lord has lined up for you to do and do it. He'll take care of all your needs. You don't have to worry about that."

Connie followed the direction of Granddad's eyes as he looked toward the barn. She knew the hay was packed in tight. The balers had cut another hundred bales today, and Granddad had helped stack it to the very top.

The pause lengthened as if Granddad was pondering every word. He went on. "Someday, Connie, you're going to meet a fine young man and marry him. Together you'll learn more and more about 'walking in truth.' "

Connie felt a warmth within her come to the surface, and her face relaxed. "Granddad, I think I've already met him," she said. Granddad just smiled.

Looking up at the blue night sky she thought, The stars have been there for millions of years doing the job they were made to do. Each person born must have a purpose too. And God directs the whole operation from way beyond the clouds. . . . No, I can't go to Houston with Dr. Yosowitz. In the distance, the lowing of a mother cow separated from her calf broke the silence.

The next morning Connie awakened to the smell of coffee brewing, and she knew Grandmother was in the kitchen cooking breakfast. Later, from the bathroom window, she could see Granddad at the barn putting out fresh hay. Mollie was nipping at the legs of some heifers who were ignoring her barked instructions to move out of the pen and into the pasture.

As Connie came into the kitchen Grandma was feeling for

bottles and boxes on the spice shelf of the cupboard. "Good morning, Grandma. Can I help you find something?"

"Oh, good morning, Connie. Yes, you can help me find the salt that one of the girls failed to put back in its place," she said smiling.

Connie found the blue box on the cabinet and put it into her hand. Watching Grandma pour the salt into the palm of her hand and sprinkle it over the scrambled eggs, Connie wondered why she never heard Grandma complain about her blindness. She knew the place for every item in the house and had always done her own housework and cooking.

"Connie, you pour us some hot coffee, and I'll get the biscuits out of the oven. Your Granddad will be along in a minute."

"Will you be lonesome when I leave today, Grandma?"

"Oh, my yes. I'll miss you. But I'll just get busy and bake a pie. Peggy Ward brought me some blackberries."

Later that day as Connie drove back to Wichita Falls, she thought of the pie simmering in the oven. Someday, she thought, I'll bake pies for Rich, keep house, and have a couple of kids—a boy and a girl.

On September 8, 1973, the morning sun rose early to hold the fog, which so often rolled in to blanket Vandenberg Air Force Base, at the ocean's edge. Across the base the bright sun rays shone through the windows of the chapel to warm the wedding guests.

Connie stood in the foyer of the church tightly clutching her dad's arm. The sounds of Purcell's "Trumpet Voluntary" set the pace for Tracy Rogers, Connie's childhood friend, college roommate, and now bridesmaid, as she made her way down the aisle. Rich's sister, Judy, gave Connie a quick hug and followed Tracy.

"Connie, you look so beautiful in our dress," Cindy whispered as she bent over to kiss her before moving into place to precede Connie and their dad down the aisle.

Connie felt a lump in her throat at Cindy's touch. As sisters they had always been close, sharing toys as children, lessons, fun, and dates in high school and college, and now a wedding gown. Suddenly she was conscious of the weight of the white satin and lace, fitted tightly in the waist and held in place with fifty-two tiny buttons down the back. She felt each link of the gold chain as it lay about her neck holding the tiny cross Rich had given her. She wondered if he had purchased it after their talk with Pastor Turner.

Her mind flashed back to that morning in the pastor's study. The quiet probing voice of the pastor had brought an embarrassed flush to her face more than once during the interview. What did a Lutheran know about "being saved?" Connie had seen Rich cross his legs and pause for a long time before explaining that he had been confirmed at Bethany Lutheran in Menlo Park at the age of twelve. Then Pastor Turner had explained that there was a difference between Jesus as Savior and Jesus as Lord. Connie had not been sure that she understood this at all.

"Rich," the pastor had said, "if you want Christ in your home, you will have to invite him there. You'll have to be the spiritual leader in your home—take the responsibility for family prayer and Bible study."

Connie remembered looking at Rich out of the corner of her eye, without raising her head. His face had been serious, and she had wondered what he was thinking. What would it be like to pray with Rich—just the two of them? Suddenly she had wanted this very much.

The last chords of the organ brought the bridesmaids' processional to a close. At the front of the church her mom stood up and glanced back just as the first notes of Wagner's "Bridal Chorus" sounded. For a brief second their eyes met. Mom, Connie thought, I love you. I'm glad we had the hour together this morning, laughing and helping each other dress.

Connie glanced quickly at her dad—tall, stately in his Air Force dress white uniform, with brightly colored ribbons and silver wings across his chest. Mom had gone ahead to the church with Rick and Cindy, leaving Dad to bring the bride. Connie suspected that Mom sensed this was a special time for the two of them.

"Keep your chin up and smile," he said to her now; "this is a happy occasion."

"Thanks, Dad," she whispered as they took their first steps down the aisle.

As they drew near the front she turned to Peg and Dick Engel,

Rich's parents, and caught their reassuring smiles. She had grown to love them these past few months. And there was Rick, her brother, winking at her, handsome in his tuxedo. Later he would walk Tracy back up the aisle.

Connie knew Rich's eyes had been on her all the way down the aisle. Now she looked just at him—this man who would soon be her husband—and saw again the gentle love in his eyes. Turning from her dad, she took Rich's arm. Together they walked up the steps to the altar.

SIX
DARE TO SEARCH

Connie sat on the edge of a lawn chair watching the morning sun creep over the edge of the wall and add sparkle to the water in the pool. At five-thirty in the morning the neighborhood was quiet except for an occasional bark of a dog and the thud of the morning paper landing on a concrete sidewalk. Mandy, the cat, returned from her morning stroll and rubbed against Connie's soft blue robe.

The house at 2409 South Clark in Tempe, Arizona, had become home for the Engels during their six months of marriage. Danny and Jan had moved out, and perfume, hair spray, and shower cap had been added to Rich's dressing room.

Feeling the touch of Rich's hands on her shoulders and his lips on her hair, Connie knew he too had had a restless night. "Rich," she said, turning slowly to look up at him, "I'm going with you. I can't stay behind while the Air Force sends you to a new assignment at Randolph."

"Of course," Rich said. "We'll fill out the joint-spouse statement today. You should be able to sell the house, pack up, and join me within three months. I'm sure the hospital at Randolph can find a place for a good GYN practitioner."

"Don't you wish you had it so easy! Get someone else to sell the house. I'm going with you. I've seen Mom pack up our family in two days so we could go with Dad."

"Your mother wasn't in the Air Force, Connie. You are, remember? You have to wait for orders."

"We'll see." Connie closed the discussion with a determined set to her chin. Rich's reassurance that she could handle the packing and sale of the house had nothing to do with the pain she felt at the thought of a separation—even a short one.

The packing began that night. "Rich, will you hang these uniforms in the front closet? We'll take them in the car with us. The rest can come with the movers. I'll take just a few kitchen items in case we can't find a house right away. Where can we put my plant friends? We can't leave them."

"Connie, I. . . . Oh, never mind. I'll call Patterson Realty tomorrow. By Saturday you should be ready to face facts."

On Friday, Connie called Rich at his office. She was jubilant. "Rich, I have my orders."

"*Your* orders? *Your* orders? Connie, mine aren't even in!"

"Check with personnel, Rich. Yours were just coming over the teletype when I picked mine up less than thirty minutes ago."

"Connie, I don't believe this. The Air Force never works that fast. Just what do your orders say?"

"That I have a wonderful husband who needs me," she teased. "I am ordered to accompany him to Randolph and replace the GYN clinician there. She just happens to be moving, leaving a vacancy which I am fully qualified to fill. Rich, I'll have my own clinic."

"You're quite a girl, you know that? But I'm not going to ask how you arranged this miracle."

"It's O.K. I just made a couple of phone calls to Randolph hospital. I knew God didn't want us separated this early in our marriage." As she hung up the phone, she could imagine Rich grinning and shaking his head.

By mid-April Connie had been promoted to the rank of Captain, and the Engels had purchased a house on Lake McQueeney, a summer resort area on the fringes of the little town of Seguin. A paved country road connected them to Randolph Air Force Base seventeen miles away, and the freeway lying on Seguin's northern limits carried the lake residents to San Antonio in twenty-five minutes.

Mandy had made friends with an orphan gray and white Persian cat who was invited to join the family. At his sister's suggestion, Rich named her Tabiatha.

Another addition, a bit more costly, came in early summer. The sixteen-foot ski boat with a 115-horsepower engine was tied at the water's edge to a deep-sunk pole in lieu of a pier. Connie's favorite time for water skiing was in the late afternoon after work. She soon learned to stay clear of the lily pads and to enjoy the freedom of half swings to the left and back to the right, gliding over waves created by a passing boat. The quiet talks with Rich as they made their way back across the lake just at dusk became important.

"Rich, are you sure Faith Lutheran is the right church for us?"

"It's the right one if they'll let you in," Rich teased.

"Any Lutheran church would be happy to get a Presbyterian. Besides, the Sagebiels like me. Half the congregation must be related to them."

"That reminds me. A captain in my office, Karl Bruce, asked me to join him and Sergeant Burwell for Bible study before work tomorrow morning. If you don't mind leaving early, I think I'd like to meet with them."

"No problem. I've got lots of things I can do in the clinic."

"Connie, what do you think about you and me studying the Bible together?"

"Fine, but it's a big book. Where do we start?"

"Let's start at the beginning—with Genesis."

Peggy Redford, the LVN who served as nurse, secretary, receptionist, and friend to Connie, got up from her desk as the last patient left the office. Walking to the inside door, she knocked lightly and opened it.

"If you've finished with the charts, Connie, I'll put them away before I leave."

"Come in, Peggy. Let's do them tomorrow. It's been a long day and both of us have husbands waiting."

"All right. I wish you could have seen Major Fox's face when he arrived this morning and I told him you two would be seeing sixty patients today."

Connie laughed, remembering how she had suggested that she

could use some help when he had phoned to notify her of his visit. Major Fox's offer to see a few extra patients while he was on base was all she needed to start the appointment calls.

"He's a great guy," she said. "I have a hard time keeping up with him. He thrives on work."

"So do you," Peggy said laughing. "You've transformed this clinic in the six months you've been here. Who else would make curtains for the windows and paint all the walls on Saturday after seeing a couple of hundred patients during the week?"

"There are times when you just can't wait for maintenance, Peggy. Did we get any word on that request for video tapes on Lamaze delivery?"

"Not yet. But the nurse at Randolph Junior High School wants to know if you can speak to her health class next week, and Dr. Smith, a local physician, would like you to speak on breast cancer detection at a meeting of the American Cancer Society."

"Let me think about both," Connie said as she gathered up her coffee thermos and purse.

"You haven't forgotten the briefing you're to give to Air Force wives at one o'clock tomorrow in the base theater, have you?"

"Oh yes, Peggy, I did forget. Exactly what did Dr. Ramos want me to cover?"

"The OB/GYN services a woman needs and those which are available here at Randolph."

"That should be simple enough. Until we get an OB/GYN doctor, there's only our clinic. If we can't handle the problem, the patient goes to Lackland. I'll get some notes together tonight."

Locking the door behind them, the two women walked down the hall together and out the front door.

Connie was becoming more and more involved in counseling. Teen-agers, young married women, older women—each group had its special problems. Dr. Yosowitz's training in evaluating each patient quickly so that routine cases could be moved more rapidly was proving to be helpful, but there were never enough hours in a day.

The hospital commander was persuaded to purchase video tape programs designed to teach about venereal diseases, birth-control methods, prenatal care, and child care. Peggy encouraged patients to make selections to meet their needs and view the educational programs while waiting for their appointments. This helped.

But Connie's concerns extended outside the clinic to the base and surrounding community, as she saw the needs for educating women to care for their bodies.

One night she got home late from a childbirth class she was teaching for expectant mothers and dads. Rich watched her as she got ready for bed. She looked tired, but she also had something on her mind.

"Rich, I have been waiting for someone to appear on the scene who is qualified to teach the Lamaze Prepared Childbirth course. Tonight I realized I've got to get certified myself. I can't wait any longer. Almost every couple there tonight would prefer that method if only there were someone to teach them. And Rich, there's another reason. When our baby is born, I want it to be a natural birth."

"You're not keeping anything from me, are you?"

Connie smiled. "No, not yet, but I should learn all I can now."

"You said that course requires many class hours, lesson plans, research. I'm not sure you can handle that right now. You've been driving yourself pretty hard these past few months."

"I don't have any choice, Rich. I've got to do it. I've been in the clinic almost a year now. It's well organized and running smoothly. I've even hung the last picture on the wall. The new carpet I managed to get was the last bit of decorating."

"You better watch out. Someone will take it all away from you."

"Not a chance."

Rich's face got serious again. "Connie," he began, "if you want to take the Lamaze course, why don't you ask for a month off? Someone from Lackland could take over for you."

"Dr. Ramos would never go for that."

"Maybe not. But he was really pleased when you were selected as the Wing's Outstanding Support Officer of the quarter. I have an idea. You're doing a lot of teaching and public speaking for the Air Force. Wouldn't the Academic Instructor Training School be a help to you?"

"Charm school? Isn't that the course for T-37 and T-38 instructors taught here at Randolph?"

"Right. I've had it. It's good. Ramos is strong on public relations. If you could convince him that you needed some training for what you're doing, he just might approve a month's TDY. You could do your lesson plans for Lamaze and practice-teach them in the school."

Connie started laughing. "Can't you just see me teaching Lamaze to a group of pilots?"

"With you teaching, they'll be interested in anything."

"The whole idea sounds like a possibility."

"I'll check into it for you."

Sergeant Burwell was a 5′11″ black man with broad shoulders and a warm personality. He loved people. Wherever he was sent by the Air Force, he established deep roots in the local community. Here, the roots were planted in a little Baptist church in a poor neighborhood with an all black membership.

For almost a year now, Connie had been meeting with Rich, Tom Burwell, and Karl Bruce for Bible study. She looked forward to Tom's weekly reports on the needs and blessings of Mount Zion Baptist Church. One morning in April, he arrived very excited about a seminar being planned for May.

"And listen, Rich," he said, "we want you to teach the men—about forty of us."

"Me?" Connie watched Rich's reaction change from surprise to elation to fear. "I'm still in Genesis," he said. "I'm not sure I could learn enough in a month."

After hours of study, digging in commentaries and dictionaries, Rich had a notebook full of notes. The night of the seminar arrived, and the Engels made their way across San Antonio just at dusk. In the parking lot Rich pulled into a place beside an old pick-up truck, and he and Connie walked across the crunching gravel toward the little white frame church. The words of "Bringing in the Sheaves" and clapping of hands to the time of the music rang across the churchyard.

Tom, dressed in suit and tie, stood outlined in the doorway, and beyond him the pews were packed. "You're here," Tom greeted them as they went up the creaky steps. "I was getting worried about you."

"We thought you said eight o'clock, Tom," Rich explained.

"Oh, that's all right. We've had a fine time singing." The hymn ended and Tom went quickly down the aisle, Connie and Rich following.

"Folks, here are my good friends, Brother and Sister Engel."

As Connie looked over the eager faces she felt self-conscious. She and Rich had dressed casually in shirts and slacks. Every woman here was in her best Sunday dress, and every man had on a coat and tie. But the friendly greetings and handshakes which followed made them feel welcome.

Later the men moved to a side room in the church, and Connie sat listening to the minister with a sinking feeling in her stomach. Rich prepared materials to support the accuracy of the Bible, she thought, but no one here doubts any part of it. The minister's simple explanations were met by "Amen," "That's right, Brother Sims," "Praise the Lord."

An hour into the meeting a tall, thin woman rushed in the back door. "Brother Sims! My toilet's running all over everything, and I can't stop it. I need help."

"I'll come with you," Brother Sims said, running down the aisle. Several women jumped up and went behind him. Within minutes the committee of Samaritans returned. Two men who had heard the commotion and met them outside agreed to handle the problem.

Later in the evening the men rejoined the women for prayer. As Rich sat down beside Connie, he handed her his notebook. "I couldn't use any of my notes," he said. She reached for his hand as the prayers began.

"Lord, I sure do thank you for my new roof. I don't have to put out buckets anymore when it rains. It sure is nice, Lord."

"Thank you, Lord, for helping me see that it wasn't doing me one bit of good to be mad at Sister Mary. She's a good woman, and she didn't mean what she said about me last week."

"Lord, I don't know what I'm going to feed my children this week if you don't help me get a job. We've been eating other people's food long enough. I want to work and earn my own living. Lord, help me find a job this week."

Connie squeezed Rich's hand. Looking up, she could see by his face that he too was learning something about a faith which offered to God a simple obedient trust in him and his Word.

A few days later, coming back across the lake in their boat just at dusk, Connie and Rich talked about the relationship with God that the people in Tom's church had.

"They have simply trusted God with their lives," Rich said.

"I agree that at some point man must make a definite choice," said Connie, "either to trust God with his life or to go his own way. But they seem to find this an easy thing to do."

The sun was just sinking below the horizon as Rich cut the engine on the boat and eased it to the side of the pier which they had recently built. Connie was quiet, sensing that Rich was deep in thought. As they tied the boat and started walking toward the house, he continued their conversation.

"Francis Schaeffer says we are to be personally directed by God's Spirit and live a supernatural life to demonstrate God's character. That's a big order."

Early in June while Rich continued to study books by Schaeffer, C. S. Lewis, and Earl Palmer in search of answers to some big questions, Connie flew to California for a week's study at an Inter-Varsity Christian Fellowship camp. Koinonia sits in a tiny valley near Watsonville, surrounded by tall pines and mountains. The gurgle of rushing water running through the camp can be heard from the cabins on a still night.

On the fourth day of Paul Byer's study of Mark, only a handful of people remained in the camp on their one free afternoon. Connie had chosen to stay behind.

As evening fell she sat with Lee Hitchcock, an Inter-Varsity staffer, at a picnic table under some pines and felt beneath her sandals the needles which had fallen from the trees. The sun, barely visible behind the mountain, cast red and orange streaks across the blue sky. Connie could feel the cool freshness of the mountain air on her face. Pulling her sweater a little closer around her shoulders, she turned to Lee.

"How long have you and Lisa been married?"

"Just three years. We married right out of college—University of California in Santa Barbara—and both went on IV staff. Lisa has a job teaching second grade next year in Georgetown, Texas. I'll join her there after camp."

"Are you going to teach too?"

"No, I'll be working in a lumberyard for awhile. I'm not sure what the Lord has in mind for us after that."

Connie reached in her sweater pocket for a Kleenex as Lee looked up to see her crying.

"Can I help, Connie?"

"I. . .I don't know how to live my life like that—trusting God so completely."

"God has given you everything you have—Rich, your family, home, profession, your life. If you hold on too tightly, you'll lose them all."

"Love is like that, isn't it?"

"Yes, and God will pour his love into your life if you will let him. Open your hands wide, Connie, and just let God take all you have into his care. You can trust him. You'll begin to know the joy of a Spirit-guided life."

In the quiet of the evening Connie put the last threads of her life in the care of the God who loved her. Shortly after returning home, she knew that Rich had made the same commitment.

On the first morning of Connie's return to work after her week in California, she was called to Dr. Ramos's office.

"Captain Engel, while you've been away I've authorized some changes which concern your clinic. The Dental Clinic needs more space. We're going to move your clinic to the north wing of the hospital and let the Dental Clinic expand into the area you now have."

Connie was stunned. She didn't know whether to laugh or cry. "I can't believe what I'm hearing. You can't be serious."

"The plan is logical. The Dental Clinic is now using two rooms across the hall from you."

"Only as a satellite clinic for fluoride treatments. The Dental Clinic is not even in this building. I had hoped someday we'd have an OB/GYN doctor in the rooms across the hall. When I came here that was the plan. I've worked hard getting my clinic set up. The north wing hasn't been redecorated in years. The paint is crumbling. The floors are impossible."

"Captain Engel, the decision is final. We don't have much money to invest, but we can make some paint available if that's what's bothering you. We're trying to run a hospital here, not win a place in *House Beautiful.*"

Connie knew the fermenting anger within her was ready to surface. So this is the way things are done in the Air Force, she thought. But it's not right.

Dr. Ramos was still talking, something about women not understanding the organization of a total program. I better get out of here, Connie thought. If I don't, I'm going to explode. She interrupted the hospital director. "Dr. Ramos, I must get back to my patients."

"Oh, yes, yes. Be prepared to move in two weeks, Captain Engel."

"Good day, sir."

"Good day, Captain."

The rest of the day went badly. Peggy responded to the news with anger, then tears, then anger again. Connie phoned Rich at his office.

"Rich, I can't believe Dr. Ramos is doing this to me. After all my work—and the curtains. What am I going to do about the curtains?"

"It's pretty rotten, Connie, I'll admit. We'll talk about it tonight, O.K.?"

That evening at home Rich cooked steaks on the grill while Connie made a Caesar salad and set a small table on the back porch overlooking the lake. The four tall kerosene flares spaced evenly around the porch kept the mosquitoes at a safe distance. During dinner Connie was quiet. She waited until Mandy and Tabiatha were devouring the last leftover bites of meat before reopening the subject of her move.

"Rich, at Inter-Varsity camp I really did tell God he could manage my life. I meant it. But why is he letting them take my clinic?"

"You were never promised an easy life. What do you want? Everything going your way? No challenges? Would that teach you patience? Kindness? Self-control?"

"Galatians 5:22."

"You've read it."

"Yes, but. . . ."

"The system is often unfair. Sometimes we have to fight to correct the wrongs. But, Connie, I don't feel this is one of those times for you."

"I know. You're right," she said, "but it's so hard not to fight back, Rich. It's so hard."

The next morning Connie and Peggy visited the old north ward. The area was worse than they had remembered. "The curtains won't fit the windows," Peggy pointed out, "and we'll never get a requisition approved to carpet this shabby floor."

Returning to their attractively decorated clinic with its soft blue carpet, they both found it hard to smile. But Connie threw herself into her work with renewed determination. I will not let this get me down, she told herself. Lord, help!

In late fall, the move completed and the old north ward now showing signs of a new face lift, Connie was selected as one of two Air Training Command Nurses of the Year. She and Rich attended a dinner given in her honor.

The time now seemed right to follow through on Rich's suggestion that she apply for a month's temporary duty assignment to attend the Academic Instructors Training School. She had completed the Lamaze seminar, but lesson plans and research were yet to be done. Rich helped her prepare the statements supporting her application for the school, and she submitted it before the Christmas holidays.

In January her request was approved. A doctor arrived from Lackland to take over the clinic for the time she would be away.

Two weeks later she sat in a room with twenty-five Air Force pilots, all men, listening to Major Jim Dawdy teach his planned introductory lesson on the T-38. "The Air Force's first supersonic undergraduate pilot training aircraft," he said, "is a two-cockpit plane. The instructor sits behind the student in a slightly elevated position. The aircraft is powered by two turbojet engines with afterburners which boost the T-38 to a speed of more than 800 miles per hour and to an altitude above 50,000 feet."

As Major Dawdy continued, Connie listened intently. She was surprised that he was holding her attention.

"The nose section of the aircraft," he said, pointing to a drawing of the fuselage, "houses the communication and navigation equipment, compass system, battery, landing light—most of the electrical equipment. Blowers cool the nose section when the aircraft is on the ground. When airborne, cooling is accomplished by overflow air. . . ."

This is interesting, Connie thought. I have not been listening to Rich because I didn't think I could understand such things. She jotted down a couple of questions to ask Major Dawdy after class. It would be fun to tell Rich what she had learned today.

The speech was drawing to a close. Major Dawdy sat down and the instructor began his critique.

"Keep your hand motions to a minimum. Movement can detract from what you are saying. Walking across the platform slowly and using the pointer to direct attention to your chart were good. These techniques relaxed the class and gave us a change."

He continued and Connie made mental notes to remember each point when her turn came.

"Captain Engel," the instructor said, "we'll hear from you tomorrow. A lesson on the Lamaze Preparation for Childbirth will be a nice diversion for us. Class dismissed."

Connie waited for Major Dawdy outside and approached him as he came out the door. "Hi. I'm Connie Engel, Major Dawdy. Your talk was really interesting, but I don't understand what difference a modified ejection seat makes."

He laughed. "Call me Jim. Your question requires a pretty complicated answer, and it would take more time than I've got right now. My wife has a birthday coming up, and I've set aside the next hour to pick up her gift, get it wrapped, and put it in the mail before the post office closes."

"That's O.K. I understand."

"Connie, I really would like to get together with you. Bonnie and I are greatly interested in natural childbirth. Maybe after your talk tomorrow we could exchange questions."

"My husband and I have a place out on Lake McQueeney. Why don't you come home with me tomorrow after class. You can meet Rich and have dinner with us."

The next night the conversation centered on flying and never got around to the Lamaze method of childbirth. Jim was an instructor at the test pilot school at Edwards Air Force Base. Connie had not realized until tonight just how eager Rich was to become a test pilot. His application was in, but he had not been selected for the last class.

"Get your master's degree behind you, Rich," Jim said. "The competition is tough. You need every break you can get."

By the end of the month's course, Connie had written a 112-page paper on Lamaze, learned to teach her lesson plans effectively, and received her certification as a natural childbirth instructor.

But she had also touched a whole new world. Maybe I'll ask Rich to teach me a few things from his flight manuals, she thought.

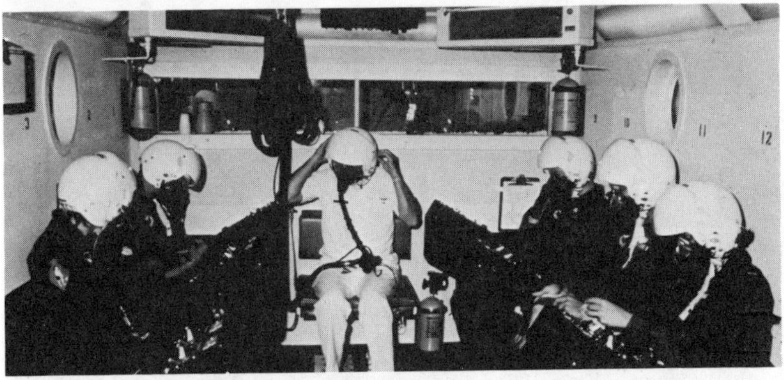

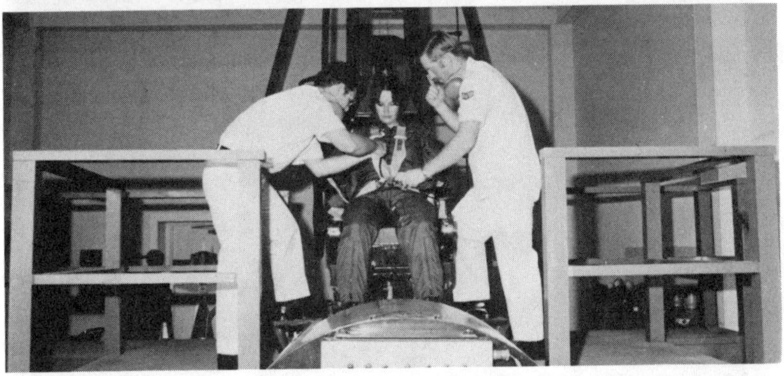

Top: Welcome to Williams, largest undergraduate pilot training base in the Air Force.

Middle: Student pilots learn about their oxygen equipment and the effects of high altitude in the altitude chamber. (The girls' first experience with the effects of helmets on hairdos!)

Bottom: Physiological training personnel help Chris Schott strap into an ejection seat trainer.

Top: Students spend many hours in the gravel pits, learning how to land from a trip down in a parachute.

Middle: Parasailing! Connie never agreed to fly without a plane.

Bottom: The T-37 students receive approximately 90 instructional hours in the side-by-side seat jet.

Top: The traditional dunking after a solo in the T-37 took on a feminine element—bubblebath!

Middle: Historically, the atmosphere in the T-38 squadron was a little more chauvinistic, and required quite an adjustment with women walking down their halls.

Bottom: The T-38 is a more challenging airplane, flying at speeds greater than 600 mph with wingtip separation of three feet. It taught Connie really to trust another person with her life.

Rich and Connie at home.

Top: Connie and her instructor, Mike O'Neill, sign out for a flight in the T-38.
Middle: At the graduation parade, Connie was glad for the USAF "front line instruction" on commands they'd forgotten to teach her in the Nurse Corps.
Bottom: Rich and Connie's dad proudly pin on her silver wings. Engraved on the back are the words "That we may touch the face of God together."

Top: Graduation awards: General Roberts, ATC Commander. Fred Clemovitz, Distinguished Graduate, Outstanding Flying Award. Mary Donahue, Academic Award. Connie Engel, Distinguished Graduate, Outstanding Officer Award, Commanders Trophy. Jerry Free, Distinguished Graduate, Outstanding Flying Award. Colonel Cooper, Wing Commander.
Middle: Susan Rogers, Connie, Mary Donahue. "You've come a long way."
Bottom: First ten women to graduate from undergraduate pilot training: Susan Rogers, Mary Donahue, Chris Schott, Vicki Crawford, Connie, Kathy La Sauce, Carol Scherer, Sandy Scott, Mary Livingston, Kathy Rambo.

Top: Rich and Connie preflight a T-38 together.
Bottom: Rich and Connie fly the T-38 together. A dream come true.

Connie.

SEVEN
SIGNAL LIGHTS

A mass of blue uniforms swarmed out of the Headquarters Building a half hour before Connie arrived to pick up Rich. As she walked into his office he was gathering up some papers on his desk.

"Hi. Are you taking work home?"

"Yes, ma'am. A new project just came out of the staff meeting with Lieutenant Colonel Hubrich," Rich said. "You will not believe what the Air Force is going to do."

"Try me."

"We're going to train ten women to be pilots. Can you imagine that? Of course it's just a test program—nothing more, but women pilots?"

"You're kidding. Oh, well, why not? The flight line already has women mechanics. If they can fix them, they can fly them," Connie said laughing.

Rich reached for his briefcase, opened it, and began loading the Air Force Tech Orders, checklists, and staff proposals he would need at home to comply with Lieutenant Colonel Hubrich's request. Rich was to evaluate the impact women would make on both the academic and physical areas of pilot training.

Connie carried her teasing further. "Rich, it's good to see you involved in this project. Because this is such a great idea, I'll even offer my assistance tonight. You might need an unprejudiced point of view."

"Great idea? Connie, where can we find ten women who have

the technical background to learn how to fly high-performance aircraft? If we did find ten, none of them would be strong enough to hold the brakes of the T-38. Women could never even pass our P.T. program, and the Air Force will never lower the standards in that area."

"Don't get so upset. I really don't know anything about the requirements, and I'm sure not interested in becoming one of your ten women."

"I'm sorry. I guess I did get a little carried away. You know, even if we managed to get ten women through pilot training, I can just see all ten getting pregnant within a year or two. Goodbye to all those taxpayer dollars poured into their training."

"At the rate young pilots are leaving the Air Force that would hardly be a change."

"True."

Connie's perception brought a smile to Rich's face as he picked up his briefcase and walked with her toward the door.

"I suggest a quiet run across the lake and a good steak before you tackle those women pilots," she said.

During the next few weeks Connie served as a sounding board for Rich as he tested and revised the recommendations concerning his area of responsibility for the women pilot project. At the same time Connie was reviewing her own career objectives. The clinic had settled once again into an orderly routine. How many times can a nurse explain birth control methods, insert IUDs, or do pap smears before they become a mechanical grind? she asked herself.

One day in late February Rich pulled his little silver sports car over to the curb in front of the base hospital. To the left of the sidewalk, Connie, wearing her navy blue trench coat over her nurse's uniform, leaned against a palm tree. She moved slowly toward the car and slid into the passenger seat.

"Hi. Let's go to Hawaii or the Virgin Islands. I need a change," she blurted out as the car shifted into gear and lunged forward.

Rich smiled as he looked at her. "When do you want to leave?"

"Oh, I don't think I really want to travel. Maybe I'll just get a master's degree in counseling, or I could go on to be a doctor. As a nurse I feel I've reached my limits except for administration, and I don't want that."

"Well, now, which will it be: Connie Engel, Ph.D. in Psychology, or Connie Engel, M.D.? Either would be useful in our family."

"You turkey. You're just interested in the social status."

"O.K. But I have a better idea. Why don't you apply for pilot training? That would be excitement enough for you. You'd be the best looking pilot in the Air Force."

"Fat chance I'd have of getting in. Besides, the Nurse Corps wouldn't let me go."

Connie's tone of voice diverted Rich's eyes from the road long enough for him to glance at her face. "You're really serious about a change, aren't you?"

"I've been doing some thinking."

"Have you been thinking about pilot training?"

"No—yes—well. . . . I'll admit listening to all the talks about flying in Academic Instructors Training School last month was interesting. But, well, flying is not my field. Anyway, a five-year commitment wouldn't fit in with a family. But then, neither would medical school."

Two days later Rich brought up the subject again. They had finished their evening meal, and Connie was flipping through an Arizona State catalogue.

"Are you looking for something in particular?" Rich asked, taking his place beside her on the couch.

"No, just something different."

"Connie, I've been doing a lot of thinking and praying the last couple of days. I'm wondering if God put you in that Academic Instructor course for a purpose. You've been restless ever since."

"Rich, if you're suggesting that I might want to be a pilot, forget it. That's a ridiculous idea. I have no technical background, and I can't run a quarter of a mile without getting cramps. I even get carsick."

"But you have been fascinated with my flight manuals, and you've been talking about flying more than you realize."

"Well, yes."

"Suppose you took the Air Force Officer Qualifying Test for pilots and scored high enough to apply for pilot training. Would you want to apply?"

"Oh, I doubt it. It might be fun for us to be in the same

profession—you know, to be able to talk the same language. But it's impossible. We would have to give up the idea of having children for at least five years. And what would our families say? I can just hear Mom now—and Peg!"

"Would you like to take the test and see what happens? You don't have to make a decision about applying for the training until we see how you do on the test."

Connie was silent for a long time. I really don't think I want to pilot a plane, she thought. I just want to learn enough to talk intelligently with Rich. There's no harm in taking the test though. No big deal. And it will make Rich happy.

The following week Connie took the test. She entered the room confidently, but when she discovered the long list of "no" answers she was having to record, she became concerned.

Answer Yes or No:

1. *Have you ever jump-started an automobile?* <u>*NO*</u>
2. *Have you ever taken a carburetor apart?* <u>*NO*</u>
3. *Have you ever assembled an electronic kit of any kind?* <u>*NO*</u>
4. *Do you have a private pilot's license?* <u>*NO*</u>

The list went on and on. Positive answers on this part of the test would be added to the total score, and her answers were all negative. On Friday afternoon a sergeant brought her test scores to her office. Connie thanked him and waited until she was alone to open the sealed envelope. As she unfolded the computer paper, her hands were shaking. She laughed at herself. I've never been so excited about a test before, she thought. Then she saw the scores. There must be some mistake, she thought. I couldn't have flunked the test. It was hard, but. . . . Her lips began to tremble. She laid the paper down on her desk. What's wrong with me? I don't want to go to pilot training. Who wants fourteen-hour days? Not me. Who cares if I didn't pass the test? I care, she thought. I care a lot. Suddenly she was crying hard. The test was so unfair. Stop this, she thought. If God had wanted me to fly, I'd have passed the test. And that's the end of it.

But it wasn't that easy. She jerked a Kleenex from the box on her desk and blew her nose. It was time to pick up Rich. I don't want to see anyone now, she thought. She called Rich and asked him to meet her outside his building.

When Rich got into the car, she handed him two sets of test scores: his, which she had taken out of his file just yesterday, and hers. Rich was still studying them as she pulled the car quickly away from the curb and headed for home. Glancing only once at Rich, she saw that there were tears on his face.

"Connie, it hurts me so much to see you hurt. You just weren't prepared for the test. You didn't have the background for it. I blame myself. I should have known. You can take the test again in six months, and we'll have you ready for it next time."

"No, I checked. The age limit is 27½. I'll be 27 in August and past the limit before the next selection board meeting. I can never take the test again."

For the next few days Connie thought of nothing except her failure to pass the test. Why can't I put it out of my mind? she wondered. Obviously I should not have taken the test. Rich's suggestion that she get a private pilot's license was flatly refused. I'll find something to interest me in the field of medicine, she thought. God has closed the door on flying.

But Rich's phone call to her on a busy Monday morning reopened her options. "Connie, the Air Staff has ruled that all women who took the AFOQT for pilots can be tested again because that test was designed primarily for males. We've got a few weeks to get you ready for the retake."

"I'm not sure I want to retake it, Rich. The door closed."

"The door just reopened. Are you giving up so easily?"

That evening Rich outlined a month's prep course for Connie. "Minority students in pilot training have a higher attrition rate than nonminority students," he explained. "Minorities often enter the program handicapped because they have not had a good technical education. The same could be true for women."

"Rich, I'm not saying I will retake the test, but just what do you think I would have to do to prepare for it?"

"Three things. First, study the book designed to prepare applicants for the test in general math and science. At the same time pursue a private pilot's license. And last, start running in the mornings before work. This will get you physically fit. Connie, if you really want to do this, I'll be with you all the way—even running beside you every morning."

"Well, those three things can't hurt me. Even if I don't get into the program, I'd sorta like to pass the test. But, if the door closes again, it's no big deal."

A cold March wind blew the drizzling rain against the sides of their faces as Connie and Rich made their way to the car early Saturday morning. Rich had insisted they keep the appointment with Jay Barry, the flying instructor he had engaged for Connie at Randolph Aero Club. But the clouds overhead were threatening.

"I think the weather will clear, but even if it doesn't, Jay will be able to show you some things about the plane," he said as they began the seventeen-mile drive to Randolph Air Force Base.

Connie watched the clouds out of the car window. As they entered the gate at Randolph, the sun peeped through the clouds and splashed on the windshield, but a few seconds later it was gone. Rich pointed out the wind sock on top of the hangar which housed the Aero Club planes. "Takeoffs will be on runway 32," he said. Connie wondered how he knew.

Two men were pushing a Cessna over to the gas tank next to the hangar as Rich parked the car. One climbed on top of the plane, reached for the gas hose, and began to fill the plane with gas. Hmmm. . .self service. Women don't do that, Connie thought. I've never even used self service at the gas station.

As they walked into the hangar, Jay met them. "Well, Connie, are you ready for your first flight?"

"Isn't the weather too bad for flying?" she asked.

"Oh, no. The weather's nice. Shouldn't be any problem at all."

"I'll leave you two, then," Rich said smiling. "I have some work to do at the office. Give me a call, Connie, when you're ready to be picked up. Have a good flight."

Before she could protest, Rich was gone, and Jay was explaining some of the rules of the club.

"We refuel our own planes. Saves us some money and keep

the cost down for club members. We sign out over here on this log," he said, walking to a large bound book and holding it up. "When you want to fly, sign up for any of the airplanes, call me and tell me when you've reserved a plane, and I'll meet you here."

Doesn't he know that this is just an orientation flight? Connie thought. I'm not sure I'm going to do this again.

"If you want to reserve a plane on the weekend, you probably should sign up early. Everyone likes to fly on weekends."

"I see."

"Today we'll take off and fly over to the south of the field—do some turns, climbs, descents—let you get a feel for the airplane. I want you to see how the plane flies, start to feel comfortable with it."

Walking out into the drizzling rain a few minutes later, Connie thought, My hair is going straight. I don't know why I bothered to curl it this morning. Girls aren't used to this kind of stuff—at least this girl isn't—and I don't think I'm going to like it.

After leading her around the plane long enough for the heavy moisture to fully penetrate her ironed jacket and slacks, Jay finally helped her into the little blue and white plane.

At least it's dry inside, Connie thought as Jay yelled, "Clear," and cranked the engine. The engine procedures held Connie's attention, but she pressed her feet to the floor as Jay put the mike to his mouth and said, "November 8112, taxi for takeoff."

"Roger, November 8112. Taxi to 32R, altimeter 30.20," the tower controller responded.

"November 8112."

Jay maneuvered the little plane through two rows of Cessnas and then past a row of T-38s. Looking out of the window at the jets, Connie thought, How could I have thought I could ever fly one of those? Absurd idea. I'll just learn all I can from this one ride so I can understand Rich's profession a little better.

Jay was talking about winds, ailerons, elevators. A T-39 landed right in front of them. Beautiful landing, Connie thought. Jay ran up the engine as he stopped just short of the runway.

"N8112 number one for takeoff."

"Roger N8112. You are cleared for takeoff. Winds are 310 at 5 knots. Have a good flight."

"Thank you. N8112, cleared for takeoff."

Jay pulled the plane onto the runway. Connie fastened her seat belt a little tighter. "I don't want to fall out if the door falls open," she explained to Jay when he saw her and smiled.

"I'll demonstrate the first takeoff," Jay said, as he added power and the plane moved down the runway and lifted into the air. As they climbed higher and higher the little plane tossed first to one side then to the other.

The weather isn't good for flying, Connie thought. I don't know Jay Barry. Maybe he isn't as good a pilot as Rich said he was. Ugh. My stomach's queasy. I'll probably be air sick. You dummy. Of course you will. Why didn't you think of that? Don't you always get car sick?

"O.K., Connie. You've got the airplane. Now fly straight and level for me."

I've got to concentrate, Connie thought as she took hold of the yoke. The plane began to climb. She pushed the yoke forward and the plane went down. Up and down. Up and down.

"How do you get this thing to level off?"

"You're doing fine. Just ease the yoke forward, just a little, very little. That's it." During the next twenty minutes Jay taught her how to turn the plane, roll to the left, level out.

"Don't let me crash this thing."

"Don't worry. You're doing great. Now let's go back to the field and try some landings."

I'm not sure I like that idea, Connie thought. Let's just land and let me out. But Jay was on the mike.

"N8112 is 7 miles south for landing."

"Roger N8112. Randolph is landing 32R. Winds are calm. Weather is 3000 broken, 6000 scattered, visibility 5 miles. Enter downwind for runway 32R. Report base."

"N8112 copies."

Jay made the first landing, but led Connie to try three with his help. When they finally came to a full stop and taxied back to the hangar, she was nervous and exhausted.

How can I tell Rich flying is an exciting sport—for someone else? she thought.

Pushing the plane over to the gas tank, Jay said, "I'll show you how to fill the tank. You can do it yourself next time."

As Connie stood watching, Jay said, "You're going to be an outstanding pilot, Connie. You handled the plane real well. You did some things that a student usually doesn't do until the third or fourth flight."

Connie reached up and touched her hair. It was wet, and straight, and cold. But the sun was shining through the clouds now, and the drizzle had stopped.

"You're going to love flying, Connie," Jay went on as he held the gas hose steady. "When do you want to fly again? Tomorrow? I think we can get an airplane after lunch. Is two o'clock O.K.?"

"Oh, oh, O.K. Sure, two o'clock is O.K. That's fine," Connie mumbled, too embarrassed to tell Jay she really didn't want to continue this nonsense. I'll never be able to fly that jet, so why should I go any further in a Cessna? she thought.

Later, as Rich and Connie drove home, he questioned her about every detail. "Did you taxi? Did you do the takeoff? Did you fly straight and level? Did you do some turns? Landings?" He didn't ask if she liked flying. Connie relaxed.

"When do you have your next lesson?"

"Tomorrow at two."

"Good. I still have some work to do at the office. Connie, I'm excited that you enjoyed your first flight so much. You're going to be a fine pilot."

It wouldn't be fair to Rich to tell him I'm quitting after the first flight, Connie thought. Just one more and I'll tell him.

But she didn't. The sun shone brightly on the day of her second flight, and her third, and her fourth. Rich and Jay's encouragement and general assumption that she would go on kept her rescheduling flights as the hours mounted up.

"Rich, it's neat that I can understand the 'angle of attack' and the operation of the wind sock. At least I'm learning a little about the aerodynamics of a plane," Connie said one day after a month of flying lessons. "I remember wondering that first day we drove up to the Aero Club hangar how you knew planes would take off on runway 32. Now I know."

In just five weeks, by utilizing her lunch hour, Connie completed the ground school which normally took fourteen weeks. The math and science lessons were being covered at a steady pace in the evenings with Rich's instruction. Best of all, the daily morning runs with Rich had given her new zest and energy for handling her busy schedule.

Retake of the AFOQT was scheduled the last week in April. As Connie walked into the examination room and was handed the test, a tremor went through her body. She opened the folder and glanced at the first page. Hmmm. . .not too bad. I know that—and that. She worked through pages one and two quickly. Page three took a little longer. The hands of the clock on the front wall moved around the circle. Connie worked on. These questions are easy, she thought. Rich was right. I needed some preparation. She finished the last page and closed the folder. As she walked out of the room, she knew she had passed the AFOQT.

A week later she landed the little blue and white Cessna at Randolph Air Force Base and taxied up to the Aero Club hangar. Jay walked over to meet her as she started pushing the plane over to the gas tank for refueling.

"Well, did you pass your check ride?"

"Of course. With you as my instructor didn't you expect me to return as a licensed pilot?"

Jay laughed and watched her as she hopped up on the plane, grabbed the hose and held it in the gas tank.

"I was just a little concerned about your having to take the check for your license with an FAA examiner at San Antonio International rather than Randolph. Too bad the guy here had to be on vacation at this time."

"No problem. I made all my mistakes before I picked up the examiner. Did you see me turn on the wrong taxiway and run head on into the T-37 when I left here?"

"I saw you."

"Did you know I forgot my instrument hood?"

"I heard you asking the controller for permission to back taxi to pick it up."

"Oh. Well, he said the examiner at International would loan me one, and he did."

"What happened to your transponder?"

"You were monitoring my radio calls?"

Jay laughed. "Sure thing."

"To be truthful, I forgot to turn it on until the controller asked about it."

The gas tank was full, and Jay helped her push the plane to its parking place and chain it down.

"Connie, the FAA examiner called me. He said you gave him a good ride. Your maneuvers, steep turns, landings—all tests were just fine. I'm proud of you."

May is a beautiful month to be in Texas. The grass and trees are green right down to the water's edge on Lake McQueeney, and the warm days are cooled slightly by gentle breezes. The sun was just sinking below the horizon as Rich and Connie headed their boat toward home after an hour of skiing.

"Rich, I've passed the AFOQT and have my pilot's license, but I'm still not sure that I should apply for pilot training in the Air Force."

"I thought we agreed not to make that decision until we fly out to California in June for Rick's wedding. By then we should have some word on my application for bootstrap to finish my master's at Arizona State, too."

"I know. We'll wait. I just can't stop thinking about what our lives would be like if I were accepted."

"We've got until July 1. Let's just not rush into this. By the time you log another forty hours on the California trip, you should know whether you want to fly Cessnas as a hobby or go on to the jets."

On June 19 Connie and Rich sat in East Valley Baptist Church in San Jose and listened as her brother, Rick, and Syndi Meadows repeated their wedding vows.

"In a good marriage," Pastor Kelly said, "each partner must give to the other not 50 percent but 100 percent. Only through unselfish sacrifice can a wife help her husband develop to his full potential—help him to become the man God intends him to be. The same thing is true in reverse. A husband's unselfish sacrifice will allow his wife to become the woman God has designed. Together the couple can learn to serve, to walk in God's truth."

Can I really ask Rich to sacrifice so much for me, Connie wondered. Flight training is twelve months of our lives—fourteen-hour days, heavy study, overnight flights. I wouldn't always be home to cook our meals, clean the house. And can we wait five years to have children? It's fine for me. I want to fly. It's the most exciting thing I've ever done. But Rich. . . .

Connie felt Rich's hand enclose hers as a group of soft voices at the front of the church began to sing, "You're a gift from heaven above/ You are meant for me and I am meant for you/ As we pledge our unending love/ Something is one that once was two. . . ."*

Two days later the Engels returned to their home on Lake McQueeney. Connie entered forty hours flight time in her log book. The decision about women's pilot training weighed heavily on both Connie and Rich. They prayed together and separately for guidance from God. They talked for hours, sometimes late into the night. The day Rich was notified that his application for bootstrap was approved, the conversation came into focus.

Connie slipped into a light blue robe following her nightly shower, let her hair down out of the shower cap, and walked into the bedroom. Rich was sitting on the edge of the bed holding a copy of his orders.

"Connie, my request to go back to Arizona State has been approved. I report to Williams Air Force Base on August 18."

Connie walked over and sat down by Rich. "My flight training would be at Williams," she said. "We could be together."

"Right."

"Rich, do you think this is God's confirmation?"

"It could be. I was reading Bob Mumford's book on guidance the other day. He talks about a harbor in Italy which can be reached only by sailing up a narrow channel between dangerous rocks. Three lights are mounted on three huge poles in the harbor. When the three lights can be lined up perfectly and seen as one light, the ship can move safely into harbor. Mumford suggests that our decisions should be based on three harbor lights: God's Word, the Holy Spirit's subjective witness to our hearts, and circumstances. We've found nothing in God's Word to prohibit your applying for this training. The Holy Spirit has given you a deep desire to fly. That is evident. Circumstances are mounting. You passed the AFOQT and got your pilot's license. We were concerned about being separated, but now I'll be at

---

*YOU'RE A GIFT by Russell Stevens. Copyright © 1973 by Lexicon Music, Inc. ASCAP. All rights reserved. International copyright secured. Used by special permission.

Arizona State with time to help you with your academics as you go through the training."

"Are you sure?"

"I'm sure."

"So am I."

Rich took both her hands in his. "Lord Jesus, right now, in this hour, we are both sure of your guidance. Help us to remember this moment every time we begin to doubt our decision. We know there are hard times ahead, but we'll reach above the clouds—with your help."

On July 22, 1976, Rich pulled the car up to their rural mailbox at the end of the road leading to their house. Connie reached in and pulled out a large brown envelope from Headquarters Air Force Military Personnel Center at Randolph Air Force Base. Rich pulled to the side of the road and waited as Connie ripped open the envelope and read aloud the letter on top.

Dear Captain Engel:
A board of officers has selected you to be one of ten active-duty female officers to participate in the test program for women pilots. You are scheduled to enter undergraduate pilot training (UPT) at Williams Air Force Base, Arizona, on 29 September 1976. Prior to entering training at Williams Air Force Base, you will attend the T-41 Flight Screening Program (FSP) at Lackland Air Force Base, Texas, beginning 23 August 1976. Successful completion of FSP is required to enter the jet phase. . . .

On August 15 Connie stood in the driveway with Rich as he prepared to leave for Arizona State University in Tempe. Every inch of the car was loaded with books, clothes, a few kitchen utensils—things he would need before the furniture arrived. The boat, secure on the trailer at the back of the car, was packed with seventy house plants which Connie hoped would survive the trip across the desert.

In two more days the packers would arrive to empty the house on Lake McQueeney of all furnishings, and the key would be handed over to the real estate agent. Connie had already begun to sort the clothing she would need for the few weeks she would be at Lackland Air Force Base. The move for her was a short one across San Antonio, but for Rich it meant two days of hard driving.

"Connie, remember the night we were very sure of our decision that you should enter pilot training. Right now I don't want to leave you. . . ."

"I know. And I'm trying hard to remember."

As Rich pulled out of the driveway Connie turned resolutely toward the house to begin the last of her packing.

EIGHT
SCREEN AT HONDO

Nine rows of T-41s, ten planes to a row, stretched out across the hot asphalt in front of the bright yellow hangar at Hondo Municipal Airport. Connie walked slowly to the end of row one, stopped, and lifted both hands to shade her eyes from the sun's glare as she watched a T-41 aircraft circle the field. The plane was coming in slowly, gliding as it banked to the right, straightened out, and dropped gently to touch the runway.

The pilot taxied the plane on the yellow line to a parking space less than a hundred feet in front of her. A young man about thirty-five, tall and thin, with fair sun-streaked hair and a narrow, sharp-boned face, climbed out of the cockpit. As he walked toward Connie he swung a small tool case lightly in his left hand and a checklist in the other. Connie saw his eyes take in her hair, flipped up just below her ears, her light pink cotton blouse and skirt, tan legs and sandals. Trickles of perspiration were running down her back. She felt hot and embarrassed. Three feet from her the man stopped and asked abruptly, "Can I help you, Miss?"

"Oh, no, sir. I was just watching you come in. That was a neat landing."

"Who are you?"

"Connie Engel. Captain Connie Engel. I'm beginning the Flight Screening Program tomorrow, Class 76-19."

"You don't say. I've been flying with the Air Force for seventeen years and never thought I'd see the day we'd let women into pilot training."

Connie felt the blush of her face and the pounding of her heart—so hard she was afraid the man would hear it.

"I believe I qualify," she managed to say.

"Do you now. Well, we'll see. We'll see." The man was nodding his head, holding his lips tightly together. Connie looked out across the rows of planes. She knew she could fly the T-41.

"Well, there's just one thing I've got to say to you, Captain," the man went on. "You'd better learn your procedures, and I mean learn them cold."

"You're an instructor?"

"Instructor and Flight Commander, John Valentine."

Before Connie had time to respond, he turned sharply and walked past her to the hangar.

Retracing the forty-five-minute drive to Medina, an annex of Lackland Air Force Base where she had checked into a BOQ early in the afternoon, Connie thought, What have I got myself into? What am I doing here instead of with Rich where I belong?

By nine o'clock that night all ten girls had arrived at Medina. Some were unpacking, but Mary Donahue and Susan Rogers sat cross-legged on Connie's bed as the three drank Cokes and talked.

"I'm not concerned about the book work," Mary said, "but the thought of being alone thousands of feet up in the air gives me goose pimples. You two have your private pilot's license. Me? I've flown coach fare on the airlines."

Susan laughed. "The flying part is easy compared to the ground school. You've sure got the jump on us, Mary, teaching mathematics at the Academy. My nurse's training didn't include heavy math—aerodynamics—not much that will help me through the academics."

"You and Connie at least had flight nurse training," Mary said.

"Those courses won't help," Connie said smiling. "We were in the other end of the airplane!"

"Well, we'll just have to help each other," said Mary. "Did you meet Chris Schott? She wants to fly fighters."

Susan scooted off the bed and stood up. "We better get some sleep. We need to be fresh and alert for General Roberts' greeting in the morning."

"And have our hair curled for the photographers," Mary laughed.

As Connie said good night and closed the door, her mind was racing ahead to a phone call to Rich. It was only 10:00 P.M. in Arizona, and he would still be studying. Her hand trembled as she dialed his number. How can I get through four more weeks without Rich? she asked herself. I want to be here, but I want to be with Rich too.

At precisely eight o'clock on Monday morning, Captain Hampton stepped inside the door of Building 147, Room 108, and called, "At-ten-tion!" The hum of ten female and four male voices ceased immediately as the students rose from their seats and stood facing the front of the room, chins up, arms at their sides.

A few minutes later General John W. Roberts, Commander of the Air Training Command, stood addressing the group, who found it hard to relax even after his warm greeting. "You will face many challenges during the next year," he said, "and winning your wings will require special effort on your part. Many people will be interested in your progress, your problems, and your achievements, as your success will have a direct and lasting impact on the Air Force."

Connie wanted to cross her legs, but she knew her chair squeaked. The room was so quiet she didn't dare move.

General Roberts went on. "Keep in mind that this program is unique. It is a test. You women will be treated no differently than the men who go through the program. You will be taught by the world's finest flight instructors, and we will expect maximum effort from each of you."

Following General Roberts' remarks and a ten-minute break, briefing on the Flight Screening Program began in earnest.

The sergeant who was to handle the physical training spoke just before lunch. "Strenuous demands on your mind and body begin today," he said, "and won't let up until you pin on your wings. I'm concerned with your physical fitness. I want you all to reach a high level of fitness before you leave here. We don't have firm standards for women yet, but you should come close to the men: one and a half miles run in twelve minutes for a starter."

Thanks, Rich, Connie said under her breath. Pulling me out of bed every morning to jog wasn't such a bad idea. She smiled to herself. I can do a mile in twelve minutes easy. A mile and a half? Hah!

By mid-afternoon the girls were being issued flight suits, boots, and jackets. Connie stood behind a curtain looking down at her men's size-36 flying suit—with thirteen zippers. The crotch hung six inches too long, and the legs crumpled in folds around her ankles. She felt the shoulder seam. About two inches over on each side.

Clumping into the outer room in the boots which were much too wide, she joined three of the girls who had just registered a complaint.

"You're not here to be beautiful. You're here to be pilots," the sergeant behind the counter was saying.

"Are these the smallest size you have?"

"Yep. The smallest."

"Do you have someone who can take them up to fit us?"

"Nope. No one I know."

"How about the Base Exchange?" one girl suggested. "Is there an alteration lady there?"

"Could be. You can ask. But it'll cost you."

Connie listened to enough of the conversation to know the man was not going to help them. She went back into the dressing room, changed her clothes, took her newly issued clothing and headed for the office of Captain Buz Hampton. "If you run into any problems today," he had said in his morning briefing, "bring them straight to me."

A sharp military type, Buz Hampton was every inch a soldier. He liked his job as Personnel Manager and Academic Instructor at Medina, but teaching a class which included ten women was not in his job description. As Connie stood facing him explaining the girls' dilemma, she could see that he was struggling to keep the half-smile on his face.

"The physical discomfort of badly fitted uniforms is sure to be a factor in effective performance," Connie continued, "even if the psychological factor is ignored."

"Captain Engel, you are certainly the right person to lead this class."

"Sir?"

"You are to be class leader."

"Just what does that mean?"

"It means that you are the top ranking officer in the class and will assume the duties outlined for you here," he said, handing her a set of official papers. "Now, as class leader, what do you suggest we do about the uniforms?"

"I suggest we find someone to alter them, perhaps through the alteration department at the Base Exchange, at no charge to the girls."

"I presume you know something about sewing."

"A little."

"Then you must know that the crotch of the flying suit cannot be raised without cutting the front zipper, and Air Force regulations state that the zipper can not be less than 29 inches in length."

"That regulation was written for men's clothing."

"True. And you women are wearing men's clothing, aren't you?"

"Captain Hampton, I. . . ."

"I know, I know, Captain Engel. I quite understand. We will get the shoulders and legs taken up a bit for you. We don't want you to be uncomfortable."

"Thank you, Captain Hampton."

"If I can be of any further help, Captain, my door is open."

As Connie left the office she wondered where she could find a sewing machine. Lord, give me patience and forgive me, she prayed, because I'm going to cut that zipper.

The next morning physical examinations started at 7:00 A.M.

"If they take any more of my blood, I'll demand credit for restocking the blood bank," Sandy Scott said, stretching out her arm.

"Would you believe I weighed in three pounds underweight?" Connie asked, patting her hips. "I've been dieting for two years to lose weight, and now I'm too skinny. Tonight I'm going to have the biggest milk shake the snack bar makes."

"Don't worry, Connie. There's plenty of room for a couple of rocks in these flight suits—no problem," said Sandy.

"My sitting height is almost too short for regulation," said Cathy Rambo. "I hope I can stretch high enough when they measure me."

"Hold your head higher or sit on a pillow," said the corpsman.

Later they sat listening to one of the doctors sum up "a few important things you women may not have read carefully in your information sheets."

"Pregnancy," he said, "will, under the provisions of AFR 106-1, disqualify you from flying duty. On the other hand, birth control pills require evaluation before you fly. Changing of the type of birth control pills or the initiation of birth control pills will cause a sixty-day grounding to insure the absence of adverse effects."

That afternoon the psychologists took over. Psychomotor tests were first. Each person was handed a paper with fifty small circles on it and told to put three pencil dots in each circle. The test was timed, and the woodpecker taps of pencils on the table made it hard to take the test seriously. The pong-type games on the computer were fun. The reading of multiple graphs was endurable. But the personality tests, so obviously designed for men, caused Connie to draw from the back of her mind the determination to keep a sense of humor. She stared at the test:

Check Two:
As a child I was most interested in becoming

1. *an engineer*
2. *a garbage collector*
3. *a rancher*
4. *a doctor*
5. *a fireman*
6. *a policeman*
7. *a banker*
8. *a scientist*
9. *a military man*
10. *a sailor*

Her eyes scanned the list. Nurse was not included. Neither was housewife or mother. She wrote in NURSE in large block letters and HOUSEWIFE underneath.

The next question was similar:

Check the activities in which you participated
when growing up:

1. *basketball*

2. *football*
3. *fishing*
4. *hunting*
5. *camping*
6. *horseback riding*
7. *Boy Scouts*
8. *repairing electrical appliances*
9. *working on car engines*
10. *mowing the lawn*

I'll get even on this one, she thought. At least once in my life I've done all of them—I even attended a Boy Scout meeting with Rick. She put a bold check by all ten.

On Wednesday morning the class was taken to the General Aviation Trainer and introduced to "Egore," the computer who runs everything. Two simulated cockpits sitting on cones occupied the center of the one-and-one-half story GAT building.

"Climb in, Captain Engel," said Nick, the white-coated man who took care of students and Egore. Connie climbed into the cockpit and heard the door slam shut.

With his mechanized voice, Egore began his instructions: "Climb to 500 feet and maintain air speed of 75 miles per hour. . . .Don't use the throttle. . . .Air speed is too high. . . . Climbing too fast. . . .Watch the ball."

Climbing out after a thirty-minute flight, Connie felt frustrated. "The air speed just doesn't correspond to the pitch," she said to Nick.

"Egore doesn't lie," he said smiling.

By Friday morning Egore had become a real challenge. Connie's assignment was to fly rectangular patterns in the airport traffic area which appeared on the screen. Egore, who was not in the best of health following an electrical storm during the night, kept going to sleep.

"Air speed is too. . .too. . .slow. Pitch is too low," he said in a monotone.

"Impossible," Connie answered as she glanced up and ran off the screen.

Egore was instantly awake. "BAD, BAD, BAD," he printed across the screen.

Keep your sense of humor, Connie told herself as she stepped out of the simulator. But Nick had not missed Egore's evaluation. "In the three years I've been here, Egore has never been so ornery," he laughed.

Leaving the building, Connie saw Randy Fields standing by her car. One of the four men assigned to the class, Randy was in the Air National Guard.

"Randy, how's your blood pressure?"

"Still up, Connie—150 over 110. I get uptight when I go to the hospital for a check. The doctor will give me the word Wednesday. I'm doing a lot of praying."

"I'll pray for you too, Randy. God may have other plans for you."

"I suppose so, but I'd sure like to fly."

"Ride with me over to class."

Captain Hampton waited until the end of the day before making announcements. "Monday you begin flying," he said. "Class as usual in the morning. Bus leaves at 11:00 A.M. for Hondo. You'll be back to Medina by 7:00. Unfortunately the Mess Hall closes at 7:00, but you can have your dinner at Lackland. The bus will take you over there."

The room was hot and Connie was tired. As she reached for her purse getting ready to leave, Captain Hampton dropped a bomb.

"One more thing," he said. "All women will get regulation haircuts before Monday morning. That means above the eyebrows and no longer than the bottom of the collar. No hairpins or rubber bands."

Every eye focused on Captain Hampton in disbelief. Connie's hair was already short, but she knew how the girls felt. She had weathered the shock when Rich had told her about the decision made earlier in his office. He had encouraged her to have her hair cut before leaving for Hondo. She rose to her feet.

"Captain Hampton, couldn't this be delayed until we get to Williams?"

"No, Captain Engel. Long hair fastened with clips could cause foreign-object damage to engines on the flight line. Or later, long hair could fall out of your helmet when you're pulling Gs. We're concerned with your safety."

Connie wondered how engine damage from hair clips or pulling Gs could possibly be factors in a T-41, but Captain Hampton's stern manner did not invite further questions. She thanked him and sat down.

"One more announcement," Captain Hampton continued. "For your own safety, women will wear no makeup while flying. That will be all for today. Dismissed."

Captain Hampton hastily left the room.

"He can't do this."

"Oh, but he can."

"How can makeup be a safety factor?"

"Your pretty face is distracting to the instructor, Sandy."

"Where can we get hair appointments before Monday?"

"Connie, as class leader, can't you *do* something—get us a little more time?"

Connie stood listening to the women. These are such minor things, she thought, but they're important. We've stepped on a production line and no longer have control—even of our bodies. How can the softness and beauty of a woman be maintained in a man's world? It won't be easy.

"Look," Connie began. "San Antonio has hundreds of beauty shops. Let's do the best we can. There are much harder things ahead. We can be sure of that. We won't have time to curl hair every night, so short hair will be a blessing."

"Connie's right. This is no big deal," Mary said. "Let's go eat."

As the class started out of the room, Connie noticed that Sandy Scott was almost in tears.

"Sandy?"

"I'm all right, Connie. I just miss David. I'd sure like to be at Fort Hood with him this weekend."

"I can understand that. It must be easier on the single gals."

That night on the phone to Rich she remembered Sandy. "It must be especially tough on the Scotts," she said, "with Sandy in the Air Force and David in the Army."

"Let's add them to our prayer list. How's Randy's blood pressure?"

"Still high. Wednesday's D-Day."

"And Buz Hampton?"

Rich listened as Connie told him about the orders to cut their hair and dispense with makeup.

"Your advice was good, Connie. It's how you feel on the inside that's important. You can help the others understand that."

"I'm trying."

"I know. You don't have to let the little things bother you. Your security is in the Lord. He'll give you the strength you need."

"Rich, I love you."

"Hey, you'll have me on the next plane to Texas."

A week later Connie was on Continental's Flight 63 headed to Williams for the weekend. As she walked into the terminal at Phoenix airport and into Rich's arms, she felt as though she had never been away.

They cooked hamburgers on a little grill outside Rich's BOQ, ate on a kitchen table by candlelight, and talked late into the night.

"Susan, Carol, Chris, and I all have light airplane experience just like the guys coming into the class from the Academy. It's hard to understand why the men were all allowed to proficiency advance and we aren't."

"Because you are the first ten women to go through flight training. If you skipped steps, the test wouldn't be valid."

"You sound like Captain Hampton," Connie said. "Whose side are you on?"

"Yours all the way."

"I'll admit Captain Hampton is a top-notch academic instructor. He must know what he's doing. And I was lucky to get Mr. Russell as my flight instructor."

"Lucky? Nothing that happens to you is luck, Connie."

"I know. God is so close sometimes—just like he's reminding me to stay calm and let him work things out."

Neither said anything for a long time. The clock struck one. "Randy washed out this week," Connie said. "He never got off the ground." The ticking of the clock was the only sound in the room. "Rich, what if I don't make it? No, don't tell me."

On Sunday afternoon Connie flew back to San Antonio and to FSP at Medina. I can take a couple more weeks, she thought. Then I'll be back with Rich at Williams. She wondered how her mother had stood the frequent separations from her dad during the many years he served in the Strategic Air Command.

Monday Connie needed all her strength to get through the day. She arrived at the track at 6:00 A.M. to discover she was the only woman running. I can't go back, she thought, as she joined the dozen athletes and they began the first quarter mile. Determined to keep up, Connie felt every muscle in her body being stretched. Four laps later she waved good-bye and kept a steady run all three blocks to the BOQ. As a warm shower relaxed her muscles, she knew she must hurry to make her first class.

On her way to Forbes Hall, a can of orange juice in her hand, she failed to notice a man in blue uniform coming across the street. As he approached the curb and turned toward her, she tried to switch the orange juice to her left hand and salute. Too late. The man stopped in front of her.

"Do women salute colonels?"

"Yes, sir. Sorry, sir."

"Who is your commanding officer, Captain?"

"Captain Hampton, sir."

"Carry on."

What else today, Lord? Connie asked later as the bus rumbled along the road on its way to Hondo. The colonel had reported to Captain Hampton her failure to render military courtesy, and the captain had called it to Connie's attention after the morning lectures. I think I missed a few things as an Air Force nurse, she thought. How can I manage so much at once?

Even before the bus stopped in front of the familiar yellow building, the women saw the television cameras and reporters.

"Oh, not today!" Connie sighed. "This is the first day of our emergency procedures."

"Yes, and Valentine's giving a formal briefing," said Mary Livingston. "Better smile. TV Ted's sharp. He'll misinterpret our enthusiasm."

Reporters were not new to the Flight Screening Program. From the day of arrival, the women had been watched by the civilian population as well as the military. Today, one reporter stayed with Connie through the briefing and all the way to the plane. Mr. Russell, her instructor, steered her through the checklist and into the cockpit in minimum time. Connie felt relieved when the plane lifted off the runway and started climbing into the sky. She knew the TV camera was on her, and she wanted to get out of sight.

Wednesday morning, September 8, the ring of the telephone awakened Connie a few minutes before the alarm clock was due to sound.

"Happy anniversary, sleepyhead."

"Rich, you're a neat husband. I think I'll keep you."

"I'm glad to hear that. Three years is pretty impressive."

"I'll miss you today."

"I have a surprise for you. Jim Dawdy is at Randolph and is going to take you out to dinner for me tonight."

"When did you talk to Jim?"

"Sunday night before he left Edwards. He and Bonnie are getting really involved in the Episcopal Church. He'll tell you about it. He is at Randolph for the Test Pilot Selection Board, but don't ask him to discuss it."

"I won't. What will you be doing tonight?"

"I'll have dinner at the Club and come back here to study a couple of hours. Call me when you get home."

"Have a happy day, Rich."

"You too."

Major Dawdy, an honest, conscientious Air Force officer, had become a good friend to the Engels since Connie met him in the Academic Instructor Training Course. As a test pilot himself, he knew the stress and strain which this job made on the minds and bodies of those who entered the program—and on their families.

Connie felt that he was studying her carefully as they made their way just at dusk along the Paseo Del Rio, a tree-lined footpath along the San Antonio River, landscaped with tropical foliage and bordered by night spots, restaurants, and shops. Choosing a table at a sidewalk cafe right at the water's edge, they sat down to have some enchiladas and tacos.

After dinner Jim sipped his coffee and listened as Connie talked about Rich, his studies at Arizona State, and her pilot training which stretched into the year ahead.

"Connie, you've been away from Rich only a few weeks. Could you hold your marriage together if you were facing a nine-month separation?"

Connie's eyes filled with tears. "Jim?"

"Of course I can't say anything, but. . . ."

"Rich has been selected for Test Pilot School! I can't believe it!"

Jim was silent, but he reached for his handkerchief and handed it across the table to Connie. She thanked him and turned away to look across the water. A small flatboat was coming down the river. Two men dressed in large sombreros and bright colored shirts were strumming guitars and singing as two small boys rowed, guiding the boat to the side of the river bank only a few yards away. The men sang, "Nita, Juanita, ask thy soul if we should part. . . ." As the strains of the song died away, several diners tossed coins to the serenaders.

Connie turned back to Jim. "The Lord has such perfect ways of working out our lives. If Rich had been selected in the last group to go to Test Pilot School, I would never have applied for pilot training. Now I'm committed. At least I'll have four months with Rich before he has to leave for Edwards. I finish here next week. Jim, Rich wanted this so much. God is good."

Connie felt as if two forces were tearing her apart inside. The thought of a long separation from Rich was almost unbearable, but her joy at his acceptance to Test Pilot School after such a long wait was greater.

"I wonder what God has in mind for you two," Jim said.

Connie didn't answer. She looked out across the water. Couples were strolling along the river's edge, laughing and talking. She missed Rich. Nine months. That's a long time, she thought.

At eleven that night, back in her room, Connie called Rich.

"Did you miss me, Rich?"

"Yes, a lot."

"Do you think we could handle a nine-month separation?"

Silence.

"Rich? You've been selected for Test Pilot School. You're going to Edwards in February."

A long silence.

"Rich, this is what you've been waiting for. Edwards isn't far from Williams. We'll keep the airlines in business. I'm so proud of you."

"Dear Jesus. Thank you for Connie, and thank you for Test Pilot School."

On September 14, her grandfather's birthday, Connie soloed the T-41. The first in her class to solo, she was met on the ground by a swarm of reporters with cameras.

NINE
SOLO THE T-37

Chandler, Tempe, and Mesa buzzed with the news. Even metropolitan Phoenix was not immune. The Phoenix *Gazette* announced with pride "Williams Makes History Again As First Women Begin Training." One local farmer, however, was heard to say, "Men in those T-38s are bad enough—but *women!*" He had watched the silver jets swoop so low the windows of his house shook.

Even the military wasn't convinced that the commanders who had opened the door wide enough to let ten women squeak through were acting in the best interests of the Air Force. Women as Air Force nurses, secretaries, even mechanics, was one thing, but women as Air Force pilots was a whole new enigma. Few members of this elite flying fraternity on whose desks the orders had landed were ready to stamp the program with their wholehearted approval.

The ten women, trying to appear oblivious to any changes or comments surrounding Class 77-08, were busy unpacking and getting settled before beginning their forty-nine weeks of 210 flying hours, 278 classroom hours, and 125 physical training hours.

Rich Engel was moving from the BOQ. He and Connie had been assigned to a three-bedroom ranch-style home in base housing. Their furniture came out of storage and for two days they unpacked boxes, taking out what they needed for their four months together and repacking a few boxes for storage in the garage. With a woman's delight Connie placed each glass and plate on the shelf.

"This silver bowl the Murphys gave us goes in the center of the table," she said. "I'll keep it filled with fresh fruit. Arizona has plenty of that."

"Here's the hanging you made to go above our bed. I'm glad we have a bed big enough for it," said Rich unfolding a four-by-six green and white felt banner on a rod with a cord on the top. Rich began to read:

I, Connie
Take thee, Richard
To be my wedded husband
To have and to hold
From this day forward
For better, for worse,
For richer, for poorer,
In sickness and in health
To love and cherish,
Till death shall part us
According to thy
Holy Ordinance

I, Richard
Take thee, Connie
To be my wedded wife,
To have and to hold
From this day forward
For better, for worse,
For richer, for poorer,
In sickness and in health
To love and cherish,
Till death shall part us
According to thy
Holy Ordinance

AMEN

Connie smiled at Rich from her perch on the kitchen cabinet. It was good to be home. The phone rang before the last box of pots and pans was unpacked. Rich watched Connie's face as she answered it and talked to Captain Lance Smith, the officer in charge of women in Undergraduate Pilot Training.

"Yes, I was notified before leaving Hondo that I would be leader of John Black flight at UPT. . . .Yes, and Kathy LaSauce will lead No Loss flight. . . .I realize the difficulties. . . .No, I haven't had much time. . . .I'll be there at 1400 hours. . . .Right, sir."

As she replaced the phone, Connie walked toward Rich. Without a word he took her in his arms and held her.

"Connie," he said after a few minutes, "I know it's a big job. I know."

"Boy, am I nervous. Academy graduates are such chauvinists.

They're going to hate me when they find out their class leader is a woman."

"True. Yes, I'm sure they will," Rich said smiling.

"You turkey. Don't kid me. This is important."

"I know, Con. You'll do O.K. Don't worry."

"Mary, Sandy, Kathy, and Carol are in my flight." Connie's mind was racing ahead. "Rich, the women don't have any clothes."

"Well, now, that's interesting."

"Hondo gave us huge flying suits, and they never did fit," she continued, not hearing Rich at all. "We've got to start all over. Captain Smith wants us at Clothing Supply this afternoon to try on boots and masks."

"Fine. You can manage that."

"But that's not all. I have an appointment with Colonel Bailey at 1400 and my uniform needs pressing. I haven't been to the commissary." She moved away from Rich and began unpacking the last box of kitchen utensils. "I thought I was going to have a few days with you, but no one ever thinks of me. There are times when I hate the Air Force."

Rich was silent. He began to gather up the paper Connie was throwing on the floor and to put it into an empty box. Connie took out the last item, a waffle iron, and placed it on the shelf.

"Rich, the Simontons gave us this waffle iron as a wedding present, remember?"

"I remember."

"The kitchen is beginning to look nice now, isn't it?"

"The whole house looks nice. That was our last box. Two days and we're all settled. How about my taking you to the Fig Tree for dinner?"

"Super. I think I'll get dressed now and make a few phone calls before I go to meet Colonel Bailey. I want to be sure the girls were all notified about this afternoon. We've got to have some decent clothes."

"That's thinking like a woman and a flight leader," Rich laughed as he gave her a quick kiss on the forehead.

By September 27 the first official report day for UPT, Connie was ready to begin training. Uniform alterations had gone well. A custom mask had been ordered because of her skinny face and long nose, the heavy boots were now feeling comfortable on her feet, and the flight suit with thirteen zippers was tucked in at all the right places.

As flight leader, she had met the previous week with a number of the men in charge of the UPT program. "Don't expect any special favors because you are women," Colonel Cooper, the Base Commander, had stressed.

Colonel Carl Bailey, Wing Commander, reinforced the statement: "You'll undergo the same rigorous training as male students. If you can't hack the program, you'll wash out."

Nevertheless, the test program was unique, as anyone reading the agenda handed to students could see.

Public Relations Agenda for Women in UPT

27 Sept. 76
1300 *Meeting with the Wing Commander in his office. Uniform will be flight suits.*
1330 *Photo session on the flight line.*
1500 *Hospital briefing.*

28 Sept. 76
0830 *Assembling of all military personnel in the O' Club ballroom.*
0900 *News conference.*
1030 *Photo session on the flight line.*
1200 *Lunch with General McBride.*

29 Sept. 76
0800–1600 *In processing for all students in Class 77-08.*

On Thursday classes began. By Saturday Connie was immersed in the T-37 academics. Because his master's program at Arizona State was demanding, Rich had kept a rigid study schedule since beginning in August. Now Connie joined him in the living room—Rich in his big easy chair and Connie on the couch, her books spread out on the coffee table. She kept flipping through manual after manual. Finally she asked, "Rich, should the Pitot heat system be on all the time during flight?"

"Only when flying in visible moisture."

"Thanks. And one more question. When static pressure is lost you can get an alternate source by breaking the glass of one of the pressure instruments. Right?"

"Break the least essential one and dump the cockpit pressure if the cockpit is pressurized."

"O.K. I've got it."

By the following Saturday questions were piling up and getting harder. "Rich, I just don't understand about forces in a turn."

"Just remember," he explained, "that when the airplane is turning, it is not in static equilibrium. An unbalanced force has to be developed to produce the acceleration of the turn. Nice and steady, the plane will produce a horizontal component of force which accelerates the aircraft to change direction. At the same time a vertical component of lift is produced which equals the weight of the plane. Here, let me draw you a picture."

Some evenings when Rich could spare time from his own studies, Mary, Sandy, Kathy, and Carol joined the tutoring sessions.

"You've got to earn the respect of the men because you are good—really good," Rich told them.

The pressure was on.

In the classroom Connie maintained a low profile while getting to know the flight one by one—Jeff Brown, Bruce Deane, John Dunstan, Jerry Free. Class meetings were held weekly. After two meetings, Connie was discouraged.

"The class is divided, Rich," she said one night after a very tiring day. "We're not pulling together. Most of the guys are friendly enough on the outside, but I know they're bugged having women in the class with photographers and reporters around all the time. I don't blame them. Class meetings are dull. None of us is relaxed. I don't know what to do."

Rich was thoughtful for a few minutes. "I have a very practical idea which you might like to try."

"Tell me."

"Bake a big batch of brownies and take them to the Friday briefing. They're sure to crumble the walls of Jericho."

"It just might work."

Before Friday the women had moved two notches up the acceptance ladder by their performance parasailing. Tech-Sergeant Daggs commanded this whole operation out on a stretch of desert which had been cleared of cacti. The soft sand had long since blown away, and Connie could feel the hard ground under her flight boots. Jerry Free had suggested the knee and elbow pads which were fitted over her flight suit.

The parachute straps attached firmly to her body and the chute lying on the ground behind her, Connie stood waiting for the truck hundreds of yards ahead of her to start moving, taking up the slack in the cord and lifting her into the air.

Why did I ever let them talk me into this? she thought. This is the dumbest thing I ever did. I never agreed to fly without a plane.

The truck started slowly.

Please, don't let them take me too high.

The truck was going faster, faster, gathering speed. The chute billowed out and lifted her into the air—higher and higher—100 feet, 200 feet, 300 feet.

I don't think I can breathe. NO! That's high enough. Lord, get my feet on the ground. Please get my feet on the ground. If I live through this, I'm quitting. Do you hear me? I'm quitting!

Slowly she was coming down. She could see the men below watching. Tom Lange, Charlie Layman, and Bill Visnovsky were walking toward her.

Look straight ahead, she thought. Relax. Relax. Her body hit the ground hard. Every bone in my body must be broken, she thought, as she lay very still, sure that she couldn't move.

"Great going, Connie."

"Not bad for a woman."

"Say, you did all right. You were forty-two seconds—longest yet."

"Let me give you a hand."

This time she was thankful for the help. Tom pulled her to her feet with one swift jerk.

"I didn't think my feet would hold me up," Connie said. She ran her hands over her body and lifted her legs one at a time. "I'm O.K. I'm O.K." Looking back down the long strip of desert, she waved her hand in the air. Sandy was next.

The following morning when John Black Flight gathered for calisthenics, two reporters from *Time* magazine were there for interviews and pictures. This was not a good day for cameramen.

And Sergeant Brown had no mercy. "Run the mile and limber up the muscles," he said, "and don't let me see anyone coming in after eight minutes."

The reporters walked over to Carol and Connie. "Let's see if you women can keep up with the men," one of them said. "The camera will be on you."

Jerry Free and Tony Gajeski moved up beside the two women as they began the run around the track. Are they really going to help us? Connie asked herself.

"Slow and easy," Jerry said. "Don't push until the last lap."

The first three laps went fine. Connie breathed easily. Her muscles were beginning to relax. But when the four were across the track from the reporters and on the last lap, she suddenly called out, "I've got a cramp. A bad one. My right thigh."

"Let's slow our pace. Keep going. How's that, Connie? Easy, easy," Jerry said.

"I can't make it."

"Yes, you can. We're almost there."

Carol was breathing hard now. "Connie, you can make it. Come on. Just a little farther."

Jerry started counting off the feet. "200, 175, 150, 125, 100, 75, 50, we're on the last 20. Start smiling. Camera is on you. Good show!"

That night Connie fell asleep on the couch before dinner. Rich carried her into the bedroom and put her to bed. Rest was more important than food.

The next afternoon at class meeting, Captain Rausch, Physiological Training officer, awarded SPLAT certificates to every member of John Black Flight. Connie's name was called last. Standing up, she said, "As certified Survivors of Parachute Landings and Techniques as well as miles and miles of jogging, I suggest we adjourn to my house and celebrate."

Not a person turned down the invitation. As Connie passed around the brownies, she knew Rich's suggestion was a good one.

"Hey, Mom Engel, you're all right."

"These are super."

Connie caught Rich's eyes across the room. She knew he'd understand how grateful she was that he'd baked the brownies the night before while she was sleeping.

Early Monday morning Connie sat in the front seat of a T-37 jet aircraft on the runway ready for takeoff. Her instructor, Captain Stephen Kaatz, was in the left seat directing her.

"Get your squawk on."

"O.K."

"Hit your clock."

"9:22."

"Hook 22 cleared for takeoff."

"Roger, Hook 22."

"O.K., Captain Engel. We're going to line up over here right at the edge of these first white lines in the middle of the runway. Pump your brakes up four or five times."

"O.K."

"Start your run up. Look outside so you know you're not creeping forward."

"O.K."

"Check your engine instruments."

"Check. Warning lights out."

"That's good. Let's go. Looking good. Don't overcontrol with your rudder. You probably won't need it anymore. Slow rotation. Looking good. Hold it. Hold it. You're clear. Don't go back down."

"O.K."

"You didn't rotate quite enough. Get your nose up. Nose up. Look at your air speed. Nose up."

"Coming up."

"All right. You made it up just fine. I'll call in. Phoenix approach. Hook 22. Passing 2000, headpin."

"Roger, Hook 22. Radar contact."

"Why don't you get your nose up? We're at 230 knots."

"O.K."

"Now follow the tech-order climb schedule. We should be at 196 knots now. Nose up a little higher. That's it."

"194, 192. . . ."

"190. Looking good. Level off at 4,500."

"O.K."

"Nose down. O.K. Looking good. You have enough air speed. You don't have to add any power. Level off here."

Now in the training area, Connie worked hard to follow Captain Kaatz's instructions. Why is my mind so slow? Think, think, she told herself. Every muscle in her body ached. After thirty minutes she was thankful to hear Captain Kaatz on the mike calling the controller.

"Albuquerque Center, Hook 22, Butte, request Sabre Recovery."

"Roger, Hook 22. Cleared Sabre. Cruise 4,000."

"Hook 22."

Only a few more minutes, Connie thought, concentrate, dummy, concentrate.

"Hook 22. Turn zero, three, zero. Cleared straight in. Runway 30L."

Connie glanced at the altimeter. 2600 feet.

"See the cattle pens?" Captain Kaatz asked. "Fly just over them. Now turn at the railroad. Watch your air speed."

"O.K."

"Roll out here. Switch to Channel 5. Start your roll out."

"2400."

"Don't overshoot it now. Roll on in."

"2000."

"Get your nose up. Looking good. Air speed good. Watch air speed. Power coming off a little bit. Looking good. Lower nose. Lower nose. That's good."

"O.K."

"Gear. . .Flaps. . .Radio call. . .Check configuration."

Five minutes later, after parking the aircraft precisely in Delta row, Connie climbed out of the plane and walked around to face Captain Kaatz who had already jumped to the ground. "Not bad for a first flight," he said as he gathered up his parachute and started across the hot parking ramp.

Connie heaved her parachute over her back and trudged along behind. Not bad, she thought, not bad. Well, I did fly that plane. I took off and landed by myself. And I'm so excited I want to jump up and down. Just what would you think if I did, Captain Kaatz? Her instructor, ten paces ahead of her, didn't look back.

Flying was underway. Every student got a hand on the throttle of the T-37 that week. And the class was pulling together at last. Laughing at each others' mistakes helped, but Connie knew that God was at work too. One day after a difficult flight with Captain

Kaatz, they both climbed out of the T-37 and started walking together back to the squadron.

"Connie, even when you goof up you're not upset. I've watched you bounce back time after time cheerful and happy."

Connie smiled, knowing that she was actually tired, hungry, and a little discouraged. "Captain Kaatz, I entered this program because I felt that it was what God wanted me to do. He knows I can't do it without him. I study hard and do the very best I can. The rest is up to him."

On another occasion Tom Lange was discouraged over an early morning flight and dreaded going up in the afternoon. Connie took a page from her small notebook and wrote on it: "Cast all your anxieties on him, for he cares for you. 1 Peter 5:7."

"Here, Tom," she said handing the note to him. "This will help."

Tom was smiling when he returned from the flight. "Thanks, Connie," he said as he passed her on his way to the flight room. "God's still up there."

By November 30 Connie had many hours of class time in Navigation, Physiology, Academics for Contact, T-37 Systems, Instrument Procedures, and Aerodynamics, and twenty-two hours with her instructor in the jet trainer.

At eleven that morning, after making three near-perfect landings with Captain Kaatz riding silently in the left seat, Connie heard him say, "O.K. Full stop this one."

This is it, Connie thought. He's going to let me solo. Pulling the plane off the runway, she rolled it to a stop.

"You're ready to take her up by yourself," Captain Kaatz said as he began unstrapping. "Usually when students solo for the first time, they hear all kinds of strange sounds that they think they have never heard before. Don't worry. Just remember all you have learned. I'm sure you can handle anything that comes up. I wouldn't send you solo if I didn't think you could. Do you have any questions?"

"No, I don't think so."

"Well, cut the right engine, and I'll get out. I'll be in the mobile unit biting my fingernails," he said laughing.

Captain Kaatz stepped out of the plane, taking his helmet and checklist with him. He leaned back over the rail into the plane and said, "I know you can do it. Good luck."

Connie had been told the day before that she would probably solo today. When she had arrived early this morning, she had watched the water flowing into the huge John Black water tank. Traditionally, students were given a dunking by their classmates after their solo flights. She had taken off and landed on this very runway many times before, but this time was different. This time she was alone. She opened her checklist and began to read down the items, performing each task as she had done so often.

Finally she pushed the two switches to start the engine and listened as it wound up to starting RPM. Sounds fine, she thought, and all instruments look good. She pushed the mike button.

"Hook 12, initial solo, taxi."

"Hook 12, taxi to runway 30. Altimeter 30.01."

"Roger, 30.01. Hook 12, initial solo."

The plane began to move slowly down the taxiway. As she turned the corner, she saw Captain Kaatz walking toward the side of the runway. The mobile unit flashed him a green light, clearance for him to cross the active runway. He'll be listening to my radio calls and watching my performance from the mobile unit, Connie reminded herself. Better make it good.

"Hook 12, initial solo, number one for takeoff," she said to the controller as she moved into position.

"Roger, Hook 12. Winds are 300 at 5 knots. Cleared for takeoff."

"Hook 12, initial solo."

As she added power and stepped hard on the brakes, she thought, I've got to hold these brakes absolutely tight. Who says women can't hold the brakes of a T-37? The aircraft didn't move an inch. She checked her engine instruments once more, and then quickly moved her feet to the rudder pedals. Moving straight down the center line, the plane surged forward. A quick glance at the panel. No red lights. Gathering speed, the aircraft screamed down the runway and lifted into the air.

"How was that, Captain Kaatz?" she asked aloud, and then laughed as she realized the mike button could have been stuck in "on" position. Now she pushed it.

"Hook 12, initial solo, request closed."

"Roger, Hook 12. Closed approved."

Watch this, Captain Kaatz, she thought. I'll show you the best closed pattern you've ever seen. Oops! Overshot the altitude by 300 feet. Oh, well, what can you expect on the first one? On downwind she did all the required checks, got the gear and flaps down, and rolled into the final turn. Lord, it's just you and me, and I'd sure like to make this a good one. Here goes!

As she rolled out on final, she remembered that she had not made a radio call.

"Hook 12, initial solo. Gear check." What a mouthful, she thought. How can I concentrate when I have to say all those words? Air speed. Watch air speed. I'm expected to land this beast. Down I go.

Crunch. What did you expect, Captain Kaatz? Grease job?

Lifting the plane back into the air, she turned crosswind to the outside pattern. I'll get a little farther away from that mobile unit, she thought. Lord, this is really neat. Just imagine! I'm flying this jet all by myself. Well, yes, I know you're helping. Sorry. Time for another radio call.

"Hook 12, initial solo. Initial." I didn't goof that one up. Now let's see if I can land this airplane. Gear. Flaps. Aim point. Air speed. Flare. Hold it. All right. How was that one? I kinda liked it myself. "Hook 12, initial solo. Request closed."

"Hook 12. Closed approved."

Gas. Oh, yes. This plane doesn't run on fumes. I better make this landing good. Don't want those guys saying a girl ran out of gas. Concentrate, Engel. You can do it. O.K. Here comes the runway.

Crunch. I'm down. Not bad, Engel. Not perfect, but not bad. The plane rolled to the hot brakes area, and Connie went through the postlanding checks. Adding power, she shut down the right engine and taxied back to the chocks. As she looked toward the building, she remembered that her dunking in the huge John Black water tank was about to take place.

Curt Gronewald, Jerry Free, and Bud Nevers were waiting as she climbed out of the plane. I wish I could at least get my boots off, Connie thought.

"Way to go, Mom."

"You did it!"

"Neat solo."

"Let me fill out my check forms and. . . ." Connie took off in a fast run, but the rest of the class quickly closed in on her. No escape.

Carrying her to the tank, laughing and joking, the guys gave her no time to remove the boots. Just before she hit the water, she noted the foaming bubbles. The men had provided a feminine touch to an old tradition. Bubblebath.

Standing, dripping, in the center of the tank, she saw Rich in the background holding her helmet and parachute. He had a big grin on his face. Traitor, she thought. Fine protector you are!

Laughing, Connie climbed out of the tank to accept the congratulations of the Wing Commander, his staff, and the publicity hounds with their cameras, who had been watching her jet solo flight—the first for an Air Force woman. They were all worried that I would damage their half-million-dollar aircraft, she thought, smiling.

After a change of clothes and a quick job with a hair dryer and curling iron which Rich thoughtfully had provided for her, Connie went into Captain Kaatz's office.

"I'm pleased," he said. "You've done well. But solo is only the first step. You have a check ride coming up next week."

More obstacles. More mountains to climb. No use dwelling on tasks already accomplished.

A few days later a letter arrived from General Mel Bowling: "Congratulations, Connie. Just don't beat Rich and me to the F-15 or you'll bring out the very worst in us." O Lord, Connie thought, you surely don't have anything like that in mind. Do you?

All the students in Connie's flight soloed in rapid succession. As solo hours accumulated on the operations scheduling board, tension mounted. Captain Kaatz, the flight commander, chose this time to introduce the fledgling pilots to the traditional John Black Horse's Award. Holding up a foot-high trophy, an orange backside of a horse with two legs mounted on a black base, he said, "This award will go to the student doing the most stupid thing during the week. Do I hear any nominations?"

Bruce Deane jumped to his feet. "I nominate Tom Fronk. I think he should stand up and tell us his own story about what happened to him over the citrus grove yesterday."

Tom got up slowly, looking rather sheepish. "Well, it was like this," he began. "I broke out of the pattern to go out to the citrus grove and couldn't get any power. I didn't know what was wrong. The plane just didn't want to go. Finally I got over the grove and circled a couple of times before Mojack gave me permission to land. Back in the pattern, I still didn't have any power. Then one of the guys in the mobile unit with his binoculars on me cracked the problem. 'Say, Hook 26,' he said, 'how long have you had your gear down?' "

Amid the laughter David Clary was trying to be heard. "We've been told that when the sun goes down," he said, "we are to land in the center of the runway rather than on the left or right side. Wednesday night Mary was lined up on the right side and her IP said, 'Mary, land on the center of the runway.' Mary responded, 'You're kidding!' The IP repeated, 'Land in the middle.' Looking down the 8,000 feet of runway stretched ahead of her, Mary put in power, picked up her nose, and headed for the 4,000 feet marker. The IP yelled, 'What are you doing?' As he grabbed the stick he heard, 'I was going to land in the middle!' I nominate Mary Livingston for the Horse's Award."

When the laughter quieted down, Mark Jacobs was standing waiting to speak. "It is my sad duty to report an infraction of the 'no solo in clouds' rule which occurred last Tuesday. I was flying with my instructor when we spotted a T-37 headed right for a whole layer of clouds. 'Hey, Albuquerque,' my IP said, 'you have a solo student up here a few miles in front of me, and it looks like he's going to need vectors to get out of the clouds.' Albuquerque didn't answer quickly enough, but we heard a feminine voice say, 'Albuquerque, this is the solo student.' The controller was very calm. 'Hook 33 solo student, where are you going?' Then we heard a tense voice. 'Hey, those clouds were closer than I thought. I'm in them!' " When the laughter stopped, Mark said, "I nominate Mom Engel for the Horse's Award."

"Say no more."

"Let's vote right now."

Mom Engel won by a landslide. As Captain Kaatz placed the Horse's Award trophy on the corner of her desk, she grinned. I'm an accepted member of this flight class now, she thought. For sure!

TEN
BEHOLD THE CLOUDS FROM A T-38

Rich finished the semester of graduate studies at Arizona State just before Christmas with five A's and one B. After a quick trip to Lompoc and Whittier, California, for holiday visits with their families, the Engels returned to Williams to prepare for another move.

By January 9, the day Rich was to leave for Test Pilot School at Edwards Air Force Base in California, all their belongings had been separated into three categories. Rich was taking the typewriter, the water bed, several boxes of books and clothing, some pictures to hang on his walls, a new bedspread, and the two cats. The couch, two swivel chairs, one set of drapes, a pair of lamps, pictures, and clothing were all in place at the new BOQ assigned to Connie at Williams. Everything else had been put in storage.

The clock struck eight as Rich picked up his Bible and *My Utmost for His Highest*, a devotional book by Oswald Chambers which he and Connie had chosen to use for 1977. "Connie, I'd like to leave by eight-thirty. That will get me to Whittier in time to drop off the cats with Mom and get a good night's sleep before going on to Edwards tomorrow morning. Come sit by me."

Seated on the couch, Rich opened his Bible to Psalm 139 and began to read. "O Lord, you have examined my heart and know everything about me. You know when I sit or stand. When far away you know my every thought. You chart the path ahead of me. . . ."

I wonder, Connie thought, if I'm really supposed to stay here. A wife's place is with her husband, isn't it? Test Pilot School is important to Rich. I should be with him.

"You both precede and follow me, and place your hand of blessing on my head. . . ."

Forgive me, Lord, Connie prayed. I know you've planned for both of us. But please go with Rich. Keep him safe. Lord, this is hard.

"If I ride the morning winds to the farthest oceans, even there your hand will guide me, your strength will support me."

Rich is going to drive away in a few minutes and I'll be here alone for nine months. I'm not sure I can. . . .

"Oswald Chambers points out that we fail to let our minds dwell on these great truths of God," Rich said. "He's right. If we could just grasp the truth in this Psalm, Connie, that God is really in control of our lives, that he has placed his hand of blessing on us, being apart for awhile wouldn't be quite so hard."

"I know. The telephone will help. And you can be sure I'll be out there every other weekend."

Rich laid the books on the table and reached for both of Connie's hands. He held them tightly in his. "Connie, let's pray. Lord," he began, "we don't need to know exactly what you're doing with our lives, or exactly where you're leading. We just need to know that you are working and are leading. We thank you that you have made this clear to us. You've promised to guide us and to meet our needs, and we believe you. I won't be here to take care of Connie. I'm depending on you for that."

"Lord, go with Rich as he drives. Prepare the way for him at Edwards. And thanks, Jesus."

"Amen."

In March Connie's class moved from John Black Flight to Beer-can Flight to begin training in the T-38. Connie was still student leader. The students found out the first day that they were faced with a whole new program.

The publications officer began the day of briefing. "Get out your T-38 Flight Manual," he said, "and let's be sure you all have current publications." Two hours later, after a page-by-page check, the standardization evaluation officer took over.

"All students will be tested weekly over flight-line procedures and general knowledge," he said. "Any student who flunks the test will be grounded, given time to study, and reevaluated by an instructor. This usually means twenty-four hours on the ground before a student is rescheduled to fly. A student who flunks three tests will be sent to the squadron check section for ground evaluation. If this testing is unsatisfactory, the next step is the faculty board. Insufficiency will wash any student out of the program."

Three more speakers followed: the flying safety officer ("Watch for civilian aircraft, and don't hog the airways"), the grade-books officer ("We graduate grade books, perfect ones"), and the computer officer ("The weight of the paperwork will equal the weight of the airplane before you graduate").

Connie watched as a nice-looking man with a high forehead and straight brown hair got up and walked to the podium. Beercan flight commander, Connie thought, and my instructor.

"Good morning. I'm Mike O'Neill. On behalf of the IPs in Beercan, I'd like to welcome you to the flight and to the squadron."

Connie kept her eyes on Captain O'Neill as he introduced the instructor pilots. I wonder how O'Neill feels about women in pilot training, she thought. I'll soon find out.

"I believe in leadership by example," he said. "That means I will never require that you have shoes shinier than mine, flight suit neater than mine, or hair shorter than mine. Enough said about 35-10."

I can manage all but the hair, Connie thought.

"Beercan by history has been the best flight in this squadron," he went on. "And we have the check-ride scores to prove it. This is because we have IPs who are willing to put in extra hours to help students. On the other hand, we've always had students who were dedicated and worked hard.

"There is no way around the fact that your class is unique. You have had women in the class since you started. This is the first time the T-38 squadron has ever had women in training, and I'm sure we will have to make some adjustments. We would like to insure that the class is handled like any other class that has gone through here, and we will expect the same standards. We're glad to be part of the first class to graduate women."

That is encouraging, Connie thought. He's sincere—and trying.

"If you study hard, put out max effort, there's not an IP in the flight who won't help you in any way he can."

Within a few days Connie was getting to know Mike O'Neill. He outlined the work for her during an initial briefing.

"Captain Engel, memorize the operating limitations and boldface emergency procedures. Memorize chapter 3 of the dash 1. Read 55-1, volume 4, and know the procedures in the local area. There are volumes of information you'll have to digest. It's like drinking out of a fire hydrant, so you must set your study priorities in the beginning. You'll be flying in a high-performance aircraft, and unlike the T-37, where the instructor sat next to you and could see everything you did, in the T-38 you're in your own cockpit with many switches not found in the back cockpit. You have to be able to handle any emergency in half the time as you'll be flying twice the speed. You've got to be sharp and think sharp."

Immediately, Connie found herself struggling with navigation, aerospace physiology, and applied aerodynamics. She wrote Rich:

I know that you know that $L{=}_s(C_L(IAS)^2S)$, *and I'm wondering why I have to work so hard to learn it. If only the information could be transferred from your head to mine. I've just spent three hours on lift and drag and symmetrical and cambered airfoils and swept wings. I finally almost understand these, and the very next chapter is* T_r *vs* T_a. *Overload! Time out!*

My ten-hour days have stretched to fourteen since I've lost my private tutor. There is too much to learn and too little time to learn it. Hardest of all is knowing I can't afford any mistakes.

Captain O'Neill was right. Flying the T-38 was much more difficult than flying the T-37. Her instructor demanded excellence. The second time he allowed her to take off, he sat silently in the backseat as the plane lifted into the air and moved swiftly into the clear Arizona sky.

"All right," Connie said, feeling proud of the accomplishment.

"Not too bad, Engel," Captain O'Neill responded. "You missed all the check speeds, had a late rotation on the stick, became airborne late, about 165 KIAS, were a little late getting the gear and flaps up, but overall, not too bad."

The plane's so fast, and my brain's so slow, Connie thought. When they reached the aerobatic practice area, Captain O'Neill took the floundering airplane and smoothed it out. He acts as if there's nothing to it, Connie thought. Someday I'll be able to do that.

"Let me show you a loop, Engel. First we want to pick up 500 knots at about 15,000 feet. Look for a mean energy level of 300 knots at 20,000 feet, or something equivalent—like 400 knots at 18,000 feet. Then put the canopy bow on the horizon, and put the throttles at Mil power. Obtain 500 knots at 15,000 feet and you're set for the maneuver. Lead the pull by about 20 knots, with approximately 20 degrees nose low, and smoothly apply stick pressure to obtain 4 to 5 Gs."

As he talked Captain O'Neill was doing the maneuver. When he pulled the 5 Gs, Connie was caught off guard. She forgot to tighten her muscles as she'd been taught in physiology. Her vision turned gray, and she felt as if she were looking down a tunnel. Everything went black for a second. Yuk, she thought. This is fun?

A half hour later, coming in for a landing, she tried hard to concentrate.

"Aim point, air speed, trim. Watch your air speed."

"I'm trying."

"Pull back, Engel. Pull back. Flare, flare!"

With all her strength she maneuvered the plane into position and landed with a rough thud on the runway. Walking back to the squadron, Captain O'Neill was silent until they were almost there.

"You need to work on your upper body strength, Captain Engel. Of course your legs must be strong too if you're going to hold the brakes. Well, you did all right." As he went through the door, he turned back to Connie and said, "You thought I was going to hold the door for you, didn't you?" Connie let it close.

Upper body strength? Opening doors? I'm doing my best, Connie thought. But, maybe he's right. I'm not very strong, and

I don't know enough. Hey, God, what's the deal? She opened the door and walked in.

After a long debriefing of all her mistakes, Captain O'Neill added, "Engel, get a good night's sleep. We're scheduled for an early one tomorrow."

Connie was not the only student feeling the pressure. In the T-37 pre-solo, the flight averaged four or five pinked rides a week. In the T-38, the number shot to two or three daily. As each student in Beercan took his initial solo flight, the whole group felt the pressure. There were many nominations for the Spilt Beer Award which had replaced the T-37 Horse's Award.

Late one afternoon as Connie was walking to her car, Carol Scherer caught up with her. "You look like you've lost your last friend, Carol," Connie said.

"Worse than that. I pinked my initial solo."

"What happened?"

"Well, I thought I did O.K., but I guess Gasser thought differently. The controller pinked me for landings."

"Carol, I'm really sorry. I'm sure you'll do O.K. tomorrow. Did you hear what Bones did? He'll beat you out for the Spilt Beer Award Friday, hands down. He was flying with the wing commander. Well, when they were taxiing back into the chocks, the colonel was filling out the 781 and not paying attention. With all the T-38s in the row facing north, Bones taxied in 180 degrees off—nose facing south. The crew chiefs were waving their arms and screaming, but Bones didn't stop until he was in perfect position—the wrong way."

Connie heard loud laughter behind her, turned around and saw Curt Gronewald, Jon Tensfeldt, and Sandy Scott who had enjoyed the whole story.

"Which one of us should nominate Bones for the Spilt Beer Award on Friday?" asked Curt.

"Hey, you guys, can I bribe you not to tell Bones I told you? I'll fix spaghetti at my place tonight."

"No bribery allowed in the Air Force, but we'll come for spaghetti. Curt and I will bring the salad," said Jon.

"O.K., turkeys. Sandy, you want to come with us to the commissary? Carol needs some cheering up. See you all at my place at 1900."

No one studied from books that night, but a lot of material got covered in conversation. Connie felt relaxed and rested when ten o'clock came and everyone said good night.

Later on the phone, Rich said, "Connie, why don't you take our boat out of storage and make it available to the students on weekends? You all need some way of relieving the pressure."

At the Friday class meeting, rules for use of the boat were established. The class agreed to take responsibility for care of the boat. Then the discovery was made that only Connie's car had a trailer hitch for towing the boat back and forth to Canyon Lake, thirty miles away.

"O.K. You can use my car," Connie said. "Just leave me some transportation—anything. If you leave early on Saturday morning, don't wake me. Get my keys the night before."

The very next morning Connie found a little blue sports car sitting in her parking space. Not a bad replacement, she thought, as she slid behind the wheel. This should be fun.

She picked up Sandy, who had been invited to study with her, and they headed to Connie's favorite retreat—the home of Donna Burke in Tempe. Donna, a widow, lived with her fifteen-year-old daughter in a big Spanish home with a swimming pool. Since Rich left in January, Connie had spent many Saturdays studying by the pool while Donna read or went about her routine.

Today Connie and Sandy drilled each other on formation flight procedures and aerodynamics. At noon, Donna came out with sandwiches and Cokes, and the three sat around a white wrought-iron table to enjoy them.

"Connie, how do you guys survive in such a hard field? I thought going back to work after Bill died last year was what I needed, but even running a simple office isn't easy. And I only work forty hours a week. I can't understand how you are able to do what you are doing."

Connie put down her sandwich. She knew Donna was not just asking the question to make conversation. The subject had come up before. She was digging for a root answer.

"When Rich was here," Connie began, "I depended on him a lot—just as you depended on Bill, Donna. After he left I knew I couldn't spend three hours on the phone with him every night. Test Pilot School is rougher than what I'm doing. Rich needs his study time—and rest. With my human resource cut off, I had to trust God. I needed help. I started getting up two hours before report time every morning, no matter what the hour. Sometimes it is two or three A.M. One hour of that time is spent with the Lord, reading his Word, talking with him. He gives me the strength I need to get through my day. I know I could not survive without his help."

Sandy had been listening intently. "Connie, I'd like to share that early morning hour with you sometime."

Connie smiled. "You're welcome anytime, Sandy."

Donna sipped her Coke, looking across the pool at the little orange tree Bill and Rich had planted three summers ago. Connie knew her loneliness.

The days slipped into summer. But there were no vacations, no days off. As the thermometer again climbed to 120 degrees, the pressure of studies and flying kept pace. The routine of local daytime flights was broken by "out-and-backs" (the one-stop, one-day flights), "cross-countries" (the minimum two-base overnight flights), and night flying.

On the solo out-and-back, Captain Ott led eleven planes, a solo student in each, from Willie to Nellis Air Force Base in Las Vegas. The controller at Albuquerque made some remark about the mother hen and her chicks as each one checked in when passing over.

"Hey, Albuquerque Center, this is Willie 12. Passing 240 for flight level 340."

"Roger, Willie 12."

"Albuquerque Center, this is Willie 34. Passing 180 for 340."

"Roger, Willie 34."

When Connie checked in, the reply came back, "Roger, Wilamina."

On many occasions her female voice created conversation on the mikes. Once, coming into March Air Force Base with Captain O'Neill, the controller asked her instructor, "Say, how many girls do you have in pilot training?"

"We have ten of them."

"Do they know how to fly?"

"Yeah, they're pretty good."

"Well, I hope your student is better than the last one who landed here. She couldn't hold her heading for anything."

Captain O'Neill cut the radio and said just to Connie, "Say, Engel, if you get off your heading one degree, you've pinked this ride."

Sometimes the controllers were sympathetic. On a flight to Luke Air Force Base to do instrument approaches, Captain O'Neill was in the front seat and Connie in the back seat. A canvas bag had been put underneath the canopy of the back seat to block her outside vision. As they approached the Luke pattern, Connie punched the mike button.

"Willie, this is. . .Luke. No, Luke, this is Willie. . . ."

"Hey, Willie 12, are you having a hard time?"

Later, returning to Willie, Connie called for a ground-controlled approach and heard another female voice reply, "Roger, Willie 12. This is Willie GCA. How do you read me?"

"Loud and clear," Connie replied, smiling as she realized another woman was going to guide her in for an instrument approach.

Captain O'Neill's voice broke in. "You women have come a long way." They had.

But Connie had a lot further to go. By August, Captain O'Neill was getting her ready for four-ship formation solo, the most exciting event of her training. After flying the formation with Captain O'Neill at the controls, she wrote Rich, "I felt actual fear as we flew precision maneuvers at high speed with wing tips only three feet apart. I really prayed. God gave me peace as he taught me what it means to trust another completely with my life. What a beautiful lesson that is."

Flight after flight Captain O'Neill trained her. "You're turning into him again, Connie. Easy, easy. All right. Now be aggressive. Move in. Easy."

The day came when he turned to Connie after printing "Excellent" on her computer grade sheet and said, "Well, Engel, you're ready for four-ship solo tomorrow. You'll have three dual ships checking you out with Captain Ott in the lead ship."

Six men and one woman. It hardly seemed fair. But that was the test. The final one of her pilot training.

The next morning Connie walked into the flight room shortly before seven for Captain Ott's briefing and sat down at a long table with the six seasoned jet pilots. A few minutes later as they made their way down the hall to the parachute shop, the gentle but pointed teasing began. "Are you sure that computer in your pretty head can push the buttons fast enough to keep your ship in line, Captain Engel?"

"O'Neill won't be there to get you out of trouble."

A smile creased Captain Ott's face and softened his label of being the instructor who taught by fear, sarcasm, and ridicule as he shut out the other men and said to Connie, "O'Neill says you're the best student he's ever taught—that you're pretty good. Today you're going to learn what flying is all about. Be tough, Engel."

When they left the parachute shop, another IP yelled, "You guys have a woman solo on your wing? Watch out. Everything you've heard about women drivers is true. See if she can hang in there."

Curt Gronewald walked to the trolley with her. A camera slung about his neck, he shouted the final word: "Watch it, Mom. It'll be on film. I expect you to say 'four' and be there."

Connie laughed and called back, "You got it."

The crew chief helped her strap in. "I'm glad my feet are planted on the ground," he said as he climbed down the ladder and out of sight. Fear of botching up the radio calls now that four ships were involved drove Connie to complete her checks in record time. When the leader's signal to start engines came, she was ready.

"Buckshot check-in."

"Two."

"Three."

"Four."

"Willie ground, AWOL 35, taxi with four."

"Roger, AWOL 35. Taxi to runway 30C. Altimeter 30.02."

"30.02, AWOL 35."

The feminine voice of the fourth check-in echoed in Connie's ears as she sat at the controls of her T-38 jet aircraft. The five-pound helmet, fastened securely with a white leather chin strap, felt heavy on her head. Her alert brown eyes peered out over the oxygen mask and quickly scanned the instrument panel from left to right. Fuel pressure, oil pressure, hydraulic pressure, exhaust gas temperature—all needles were parallel. Her eyes leaped to the yellow line on the taxi strip stretching straight ahead, flat, bright. The fingers on her right hand tightened in a firm grip on the stick, blocking an impulse to tremble. The stick felt damp, and she knew her hand was sweating.

As she moved into position to taxi the empty back seat confirmed her command of the million-dollar aircraft. Captain

O'Neill, who had swiftly corrected her mistakes and just as swiftly pulled her out of precarious situations, was not there. Often he had reminded Connie that a T-38 could travel faster than the speed of sound and was anxious to do so at the flick of a pilot's finger.

"Willie tower, AWOL 35 with four, number one for takeoff."

"Roger, AWOL 35 is cleared for takeoff. Winds 300 at 10 knots. Frequency changes approved."

"Roger, AWOL 35. Buckshot go three."

"Two."

"Three."

"Four."

Out of the corner of her left eye, Connie watched the lead T-38 supersonic jet pull into takeoff position. Seconds later, Buckshot Three lined up to form the left leg of a triangle, and Buckshot Two eased into position to complete the right leg. Her ship still rolling, Connie's eyes caught the glare of sunlight resting on the helmet of the lead pilot. When it was covered by the helmet of Buckshot Two, she was in position.

"Buckshot, run 'em up."

"Two."

"Three."

"Four."

This time Connie's voice over the radio frequency echoed hollow in her ears, as if the muscles around her lungs were holding taut and not pushing up the air to make the sounds. Perched like silver bullets ready to be fired, the noses of the four jets were aimed for accurate shots down the strip of gray concrete. Seconds later they were rolling.

"Phoenix, AWOL 35, flight of four, passing 1,500 for flight level 270."

"Roger, AWOL 35. Radar contact."

In the air, the four ships moved quickly to the assigned formation area. At first the maneuvering was easy. Nothing new. Connie's confidence grew. "How's that, Captain O'Neill?" She tossed the question to the empty back seat, feeling her instructor would be proud of her.

Position change. Now as number two, flying on Captain Ott's wing, she felt a particular need to stick in there. Maneuvers were getting harder and harder. She worked harder and harder. She checked her G-meter at 4Gs. You turkey! You're working me to death! she thought, but smiled as Captain O'Neill's voice rang in her mind. "Hang in there, Engel. You're tough." A few times she was glad her instructor couldn't see her. "Get in there, Engel. Don't accept this position. Are you out of position?" Strange how his commands directed her decisions.

Tight in formation, the four aircraft eased into a tunnel of white billowing clouds, dimming the glare of the sunlight to a soft glow. The dew held by the clouds before being dropped to earth gave the whiteness a mellow silver sheen. Connie knew she would never again see anything so beautiful, the work of God. Instantly, she had the feeling of being enveloped in God's love, immersed in gentleness, touched by the Creator of all the universe. She was aware of the power of love, penetrating the depths of her soul. Something wonderful is happening, she thought. The nearness of a loving heavenly Father held her in awe, breathless. Tears of pure delight ran down her face and mingled with the sweat of her flight suit.

The sun burst forth as the formation came out of the warm glow of the clouds, and Captain Ott called for clearance to go home. Four ships abreast, tucked in tight, the flight home began. Nearing Williams, the four pitched out to land.

"Buckshot, lead, gear check."

"Two, solo, gear check."

"Three, gear check."

"Four, gear check."

The feminine voice sounded confident, peaceful, sure of God's provision. Walking back to the chute shop with the six men, Connie smiled inwardly, listening to their pilot chatter. In the flight room the talk of this turn or that turn was drowned in congratulations from Captain Ott as he printed a big "E" for excellent beside "Four-ship solo" on her grade sheet.

"Thank you, Lord," whispered Connie as she made a hasty retreat to her car and headed toward the BOQ to change her clothes for her flight to California. The calm peace within her was slowly giving way to an overflowing joy. She wanted to be with Rich.

The Engels' conversation over the weekend centered on their assignments. "Captain O'Neill hinted that I might be assigned to Willie as a T-38 instructor, Rich."

"Would you like that?"

"Only if you're there." She was quiet for a moment. "As a test pilot you're more likely to remain at Edwards. Right?"

"Right. Or perhaps Eglin Air Force Base in Florida. You could be assigned as a chase pilot in the T-38 at either base. What would you think of that?"

"That would be great. I don't care where we go or what I do if we're together."

"We'll have to trust the Lord for that."

Sunday night as Connie drove through the front gate at Willie, she knew Beercan Flight had been working some mischief over the weekend. On the "Welcome to Williams AFB" sign the names of the Wing Commander and Base Commander had been replaced. She stared in disbelief.

WELCOME TO WILLIAMS AFB

Flight Commander	*Class Commander*
Captain Pappy O'Neill	*Captain Mom Engel*

A few blocks further on the huge Beercan flag flew proudly from the top of an abandoned boom bucket. The sign on the front gate was quietly changed the following morning, but the Beercan flag appeared at various places on base during the next couple of weeks.

At the last Commander's Call in the Base Theater just before graduation, Lieutenant Colonel Bartlett entered and the students stood at attention. As he approached the stage the curtain slowly began to close. Sewed to the heavy drape was the proverbial Beercan flag. As laughter swept the theater, Lieutenant Colonel Bartlett scanned the audience of two hundred students and said, "Captain Engel, where are you?"

During the final week, students in Beercan Flight showed their appreciation to their instructors by arriving en masse, armed with food and drinks, at the home of Captain O'Neill and then at the home of Captain Ott. The instructors had become more than respected pilots. They had become friends.

The class had pulled together, and Connie knew that she too had won their respect when she found a ten-speed bicycle in her parking place at the BOQ, left as replacement for her car and boat.

To add to the excitement, class assignments came in from the Personnel Center at Randolph Air Force Base, Headquarters for Air Training Command. The women were particularly concerned. Mary Donahue and Sandy Scott got KC-135 Stratotankers, Kathy LaSauce and Kathy Rambo the C-141 Starlifters, Susan Rogers and Chris Schott the C-9 Aerovac planes. Carol Scherer was assigned to the WC-130, Vicki Crawford to the T-43, Mary Livingston to the T-37. Connie was assigned to Williams Air Force Base as a T-38 instructor pilot.

Following graduation, she was ordered to Instructor Pilot Training at Randolph Air Force Base. The new year, 1978, would begin a three-year assignment for her at Willie. As Rich would not complete Test Pilot School at Edwards until December 18, 1977, they would not know until then what his assignment would be.

"There is a rumor of a test pilot being assigned to the Human Resources Laboratory at Willie on a new program, Connie. It would have to be a miracle if I got the assignment."

"That's God's specialty," Connie said.

"Right. I'm sure glad he's the Commander with the final word on our assignments. Don't be concerned no matter what comes out of this. God even has veto power."

September is always a beautiful month in Arizona. The wind is at rest, allowing the dust to remain on the ground and the air clear. The stars are bright over the desert, and the sun, as it rises and sets, throws a blaze of color across the blue sky.

On September 2, graduation day for Class 77-08, the strains of the marching band were carried much farther than the reviewing stand. As Air Force dignitaries, instructors, parents, and friends watched the formal salute, Connie stood at attention with Beercan flight. She had learned very little about parade procedures inher six years in the service. As a nurse, she had marched only once. But with forty-six Academy graduates in her flight, she wasn't worried. They would do the right thing no matter what command she called, and Craig Kinney standing behind her was whispering the commands to her seconds before she needed them.

Connie knew Rich was seated in the stands with their parents. Peggy and Dick Engel had driven over from Whittier the day before, and Judy, Rich's sister, had arrived an hour later. Connie's parents had flown in from Los Angeles on a late Thursday afternoon flight.

The parade lasted less than thirty minutes, and the crowd moved to the Base Chapel for the graduation address. Colonel Larry Cooper, Commander of the 82nd Flying Training Wing, introduced General Roberts.

Connie recalled the address the General had given at Flight Screening Program at Randolph a year ago. He was talking now about tradition.

"The military has many traditions," he said. "Some have deep roots. Some are being challenged by technology, some by sociological changes. Today marks the end of the tradition of an all-male flying force. Today we are standing on the threshold of a new tradition—the tradition of men and women flying together in defense of freedom. This is an historic event."

Rich and I are the first married pilots in the history of the United States Air Force, Connie thought. In an hour we will be answering questions for newspaper reporters and for a CBS television interview. I'm glad Rich is with me.

General Roberts' address was over, and Colonel Mehaffey, Deputy Commander for Operations, was presenting awards. "The Officer Training Award goes to the student who has exhibited high qualities of military bearing and leadership," he said. "Will Captain Connie Engel please come forward?"

Connie rose, went forward, and was given a beautiful trophy with her name inscribed on it. This award traditionally went to the class leader.

Mary Donahue earned the Academic Award. She had missed only one of 365 questions over a year of examinations. The flying award was shared by Jerry Free and Fred Clemovitz.

Again Connie heard her name being called and went forward to receive one of three Distinguished Graduate awards. As she returned to her seat, she prayed silently. Lord, this is really a bonus. It's hard to believe this has happened to me, but thank you.

General Roberts had risen and was explaining the last award, the Commander's Trophy. "This award is the most coveted award in any class," he said. "It is given for overall excellence in flying, academics, and leadership. In a class of ten outstanding women and thirty-six outstanding men, this award is especially significant. It goes to Captain Connie Engel."

Connie could not move. This isn't possible, she thought. Lord, is it true?

As she finally started down the aisle toward General Roberts, tall and dignified in his Air Force uniform, her knees felt like sponges and tears ran down her face. I'm sorry, Lord, and General Roberts. I know I'm an Air Force officer and a pilot, but I am also a woman.

Later on the lawn of the chapel, Rich pinned on her silver wings. Connie knew how very much she loved him when she read the words engraved on the back of the wings:

That we may slip the surly bonds, and touch the face of God together.

Connie studied Rich's face as he struggled with the clasp. This was his day too, and they shared the silver wings. Our lives are so intricately entwined, Connie thought, that everything we do, even the smallest thing, deeply affects the other. Rich knows how I feel. He knows.

The wings in place, Rich gave the new Air Force pilot a big hug. As they turned and started walking toward the waiting television camera crew, Connie's dad stopped them.

"Connie," he said, handing her another trophy, "I made you this award to help you remember that God's help and protection are always with you—even in the clouds. He's the master Pilot."

Inlaid on the polished oak was a pair of wings with a white dove above them. Below on a brass plaque were the words of Psalm 63:7.

Because thou hast been my help, therefore in the shadow of thy wings will I rejoice.

Connie looked up at her dad, now retired from the Air Force. There was strength in his face, and love. "Thanks, Dad," she said. "I'll remember."

ELEVEN
JOY BEYOND THE CLOUDS

Rich smiled at Connie, sitting beside him on Delta's Flight 112. They would be landing in Montgomery in less than an hour, where Cindy and Ed would meet them to begin the seventy-mile drive to Enterprise near Fort Rucker. It was Christmas Eve.

"Rich, it's pretty neat, isn't it?"

"What?"

"That we're both stationed at Willie again, and that we're both pilots in the Air Force, and that everything worked out just right."

"Yeah, it's amazing. I'll admit I was a little nervous before I got the orders in my hands. Human Resources Lab, working with the advanced simulator for pilot training, doing research like I was taught at Test Pilot School—wow!"

"Are you really happy with the assignment? I mean, you won't be flying test airplanes, or chasing the F-16, and you're not likely to be swooped up by NASA for the next space shuttle. Rich, does it matter that you were assigned with me and not me with you?"

Rich looked out the window of the plane for a few minutes before answering. The sky was bright with the sun's rays, filtering through the clouds spaced at intervals in the blue sky.

"Connie, I worked out that problem a long time ago. After being introduced as 'Captain Engel's husband' a few times, I had to face it. I've come to realize that my worth as an individual does not depend on what others think of me. In God's sight I have great value. He is the one I must please. That means doing my very best in whatever job I have. If God chose to station me with you and not you with me, so be it. I know he has a purpose for us. If he wants us at Willie, that's where we'll go."

Rich was thoughtful for a moment, and Connie waited for him to go on.

"Back to the original question," he said smiling. "I am excited about the Research Lab assignment. Besides, what better place could I be than near Arizona State University to finish my master's degree? And putting you in an environment that will give you a little experience and confidence in your flying was not a bad idea. You might have been eaten alive as a chase pilot at Edwards with all those golden pilots out there."

"Wait a minute, you turkey. You think I couldn't hack it? Well, you're probably right. I get nervous just thinking about it."

The plane was coming in on final approach at Montgomery Airport. Through the window Connie could see the runway below, buildings, cars, trees, people. The big plane straightened out and descended with only a slight dipping of wings from side to side. The roar of the engines' reverse thrust and brakes grinding the tires in the concrete vibrated through the cabin. I wonder if I could hold those brakes, Connie thought. Rolling to a stop, the huge aircraft began moving more slowly to the taxi strip, making its way to the terminal. Connie imagined the conversation on the radio with the ground controller, now directing the pilot. She turned from the window to help Rich gather up the Christmas packages under their feet. Cindy and Ed would be waiting. They had bought their first house eight months ago, and had insisted the whole family come to Alabama for Christmas.

Two hours later Ed turned the car onto Seminole Street in Enterprise and headed toward the stained wood two-story house on the corner. Connie noticed an overturned tricycle in the yard across the street, and saw a group of children playing in the cul-de-sac a half block away.

Cindy was describing the occupants of each house. "Alice and Ken live there. They have three kids who play with our two. The Warrens are in the brick house. They're expecting their second child next month."

Ed had pulled into the driveway. As they climbed out of the car, Cindy was still talking. "Last night when we got back from

Montgomery with Mom, Dad, Syndi, and Rick, my neighbors had dinner in the oven waiting for us—fried chicken, butter beans, corn bread. I love living here."

"Don't get too attached," Ed said, putting his arm around her waist as they walked toward the house. "You're an Army wife."

"What would it be like to have time to really get to know our neighbors, Rich?" Connie asked him quietly as they reached the porch.

Rich laughed. "We'd need a lot more hours in our day."

Connie's mom and dad opened the front door, but Hil and Mark, ages seven and three, managed to get the first hugs. Syndi and Rick waited their turns. The house smelled of fresh-cut pine branches, bayberry candles, and baked ham. Connie could see traces of Cindy's preparation for the holidays in every room: the manger scene on the coffee table, the well-trimmed tree, holly and mistletoe in the doorways.

Within the hour dinner was on the table. It was Hil's turn to ask the blessing. Now he glanced around the table to see if everyone was ready. At a signal from his father, he bowed his head and began.

"God, today is a happy day. Mom and Dad and Mark and I are glad everybody is at our house—Grandmommie and Grandad. . . ."

"And Rick and Syndi and Connie and Rich," Mark chimed in.

Hil sighed and went on. "Thank you for the food—especially the pecan pie. Amen."

That evening when the family gathered in the living room for the traditional reading of the Christmas story, Connie sat with Rich on the floor in front of the fireplace. Hil snuggled close to Rich, who had earlier won his admiration by explaining why his electronic buzzer wouldn't work and then by helping him repair it. Mark climbed into Connie's lap. The children's eyes and chatter kept returning to the Christmas tree in the corner of the room. The bright-colored lights cast multi-hued designs on the gingerbread-men which Cindy and the boys had baked and hung with red ribbons on the green branches. Presents were stacked three-deep under the tree.

Ed passed the Bible to Dad who smiled, adjusted his glasses, and began to read from Luke 2. "And it came to pass in those days, that there went out a decree from Caesar Augustus, that all the world should be taxed. . . .And all went to be taxed. . . ."

As the familiar words filled the room Connie looked around her. Mom was seated on the couch beside Dad, her arm linked through his. Mom will miss her former students dropping by this Christmas, home for the holidays, Connie thought. Mom and Dad have been married for more than thirty years, and Mom has managed a home and has taught for half that time.

"And there were in the same country shepherds abiding in the field, keeping watch over their flock by night. And, lo, the angel of the Lord came. . . ."

Mark was getting sleepy, and Connie felt the weight of his head on her arm. Her eyes drifted to Rick and Syndi, squeezed tight into Ed's big leather chair. Syndi was an answer to prayer. Connie smiled, remembering. In Rick's junior year in high school, Mom had prayed he would meet a nice girl at Christian camp that summer. He came home with Syndi's picture. During Rick's first year at Texas Tech they became engaged and were married in June. Mom thought they were too young, but God sure knew what he was doing.

"And they came with haste, and found Mary, and Joseph, and the babe lying in a manger."

Cindy smiled at Connie across the room and motioned her to look at Hil whose head was now resting on Rich's shoulder. He was blinking his eyes to stay awake. What a blessing children are, Connie thought, but what a hard job—being responsible for their existence, guiding them, teaching them, feeling deeply their hurts, fears, failures. And what if you lose them or fail them?

". . .a light to lighten the Gentiles, and the glory of thy people Israel."

Dad handed the Bible back to Ed. "What a wonderful way for God to redeem us and teach us about himself," Dad said, "by becoming one of us."

Ed and Cindy got up and each carried a child upstairs. When they returned to the living room, Cindy said, "I think I better replace a few gingerbread men. The boys have been nibbling on those at the back of the tree." She went to the kitchen and came back carrying a big tray of cookies. Ed got up to help her. Connie watched them talking and laughing quietly about something the boys did during the day.

"Connie," Ed said, "you better get prepared. Cindy has asked some of her friends over to meet you Wednesday."

"I hope you don't mind, Connie. The girls have all been following your career—saving clippings for me. Bonnie even told me when you were to be on television. We were talking about you just last week at Jenny's house, packing Christmas boxes."

The fire was burning low. Rich got up and added another log.

"Connie," Cindy went on, "I've often wondered how you could take on such heavy responsibility, setting an example, instructing students. Some of them won't make it, and you'll be hurt more than they will."

"I was just thinking what a responsible job you have, Cindy, and what a good job you and Ed are doing with the children."

"There is never a dull moment in my life, but it's great fun," Cindy said, laughing.

Connie turned to watch the fire, now burning brightly. She moved close to Rich, and he put his arm around her. Granddad called it "walking in truth," she thought. It's moving through life according to God's unique plan for each of us. It's giving of ourselves and doing our best in whatever job we have. In God's sight every job carries great responsibility.

Connie remembered that Christmas Eve and the talk by the fire many times during the next few months as the publicity mounted. She and Rich were the first married pilots in the history of the Air Force, and on rare occasions they flew together in the T-38. One time Rich called ahead for BOQ reservations at Travis Air Force Base, California.

"Name?"

"Engel, Richard."

"Other pilot?"

"Engel, Connie."

"Oh, yeah? You married?"

"Right."

"Yes, I guess, I mean—well, it is a little unusual. We've never had this before."

When the Engels landed and walked into base operations, the young airman on duty read their name tags.

"Engel and Engel. You guys married?"

"Yes, we are," Rich answered grinning.

"You're kidding. You both fly?"

"Yes, we're both pilots."

"What do you fly?"

"The T-38."

"Oh, yeah? I should have had the photographer here. This is a first."

Fifteen minutes later they walked into the BOQ office and were met by a dozen guys—and the base photographer.

"Here they are! This is the couple I told you about—husband and wife pilots. Can you imagine that?"

The Engels bought a home in Mesa, not far from the base. "We'd like to open our home to men and women going through pilot training here at Williams," Rich told Larry Stair, the base chaplain. "They can come, swim in the pool, read or study in a room we've fixed up for that purpose."

"And we want to have some talk sessions too," Connie said. "We've struggled with lots of the big questions in the last few years. We're planning open discussions on Sunday afternoons for anyone who is interested. By sharing questions, ideas, experiences, maybe we'll come up with some answers."

"Maybe you will," the chaplain said, smiling.

Several weeks later, twenty Air Force men and women gathered at the Engels' home. They had been swimming for a couple of hours, but now the strumming of a guitar drew them inside. Jerry Free and his new wife, Joyce, were sitting on the floor in the living room, Joyce playing the guitar and both singing an Andraé Crouch song.

You may ask me why I serve the Lord—
Is it for heaven's gain?
Or to walk those mighty streets of gold
And to hear the angels sing?

Connie quietly walked into the room from the kitchen. She had arranged the last slice of tomato on a tray for serving later with the hamburgers.

But if heaven were never promised to me....
If there were never any streets of gold,
*It's been worth it just having the Lord in my life....**

Jerry and Joyce Free, Bob and Margie Frolich, Phil Julien and Vince Bonasso had joined the Engels in planning for the Sunday afternoon open houses. Today Rich was leading a discussion on John 5:24 and fear of death.

"I didn't think I'd admit this to anyone," a young student, Howie Donahoe said, "but I really panicked this week on an initial T-38 flight with my instructor. He did the maneuvers so quickly my head was spinning. I had a real fear of dying."

"I can identify with that," someone in the back of the room said.

The discussion went on, some saying they had never been afraid of death, most admitting it was a reality in their lives.

"Death is a part of life," Connie said from her place by the door. "An old man once said that to me. I know now that we have to experience death before we can know life."

That night after everyone had gone, Connie was unusually quiet. When she wandered out by the pool just after dark, Rich followed.

"Hey, beautiful. What are you thinking about?"

"Oh, our discussion this afternoon. I'm taking up my first student tomorrow."

"Are you nervous?"

"I was thinking about Jim Alexander." The night was very still. Connie looked up at the clear Arizona sky, filled with bright stars. "Rich, are you afraid for me?"

"No, I don't think so. The only fear I had for you during your training was that you might not achieve your goals. I knew that would hurt you very much, and I didn't want you to experience that hurt. You're an instructor now. You have a lot to give your students. Being unwilling to share our lives with others, to teach them what we have learned, would be the worst kind of death. I have no fear for you."

---

*IF HEAVEN NEVER WAS PROMISED TO ME by Andraé Crouch. Copyright © 1973 by Lexicon Music, Inc. ASCAP. All rights reserved. International copyright secured. Used by special permission.

For three weeks students in Beercan Flight had been briefed, taught, and drilled. Today many of them were scheduled for initial flights with their instructors. At precisely 4:30 A.M., Connie walked into the briefing room with eight other instructor pilots. The class leader called the students to attention and saluted Captain O'Neill, still the flight commander.

"All students are present and ready for the briefing, sir."

Captain O'Neill replied, "Take your seats."

Students turned to their instructors, saluted them, and took their seats at tables across from them. Lieutenant Dan Schmid and Lieutenant Greg Robinson sat across from Connie. My students, Connie thought, young, alert, expectant. Air Academy graduates, both looked the part. Lieutenant Schmid with his brown hair and hazel eyes was studying Connie's face. He's wondering why I'm here, Connie thought, the only woman in the flight room. Well, it's a long story. Someday I may tell him.

Lieutenant Robinson, whose dark eyes were set in a round suntanned face, didn't have a hair on his head out of place. The light gave it a coal black sheen. He was sitting very straight in his chair, his eyes on Captain O'Neill standing at the podium. Connie listened.

"Get your time hacks on the wall. The date is One March. Flying status is Alpha."

The pilots set their watches. Captain O'Neill continued. "Surface winds are 310 at 8 knots. All the navigation aids are in. Willy is on runway 30, Tucson is on 11, Davis Montham 12, and Luke 21. Areas are clear. Should be good flying. Some of you will go up in the white rocket for the first time today. I'm sure you're excited. Listen to your IPs and take your time strapping in the machine. Fly safe."

Connie watched him turn and walk into his office, his new student behind him. She turned to her students.

"Lieutenant Schmid, I see you're scheduled for link trainer. Lieutenant Robinson and I have a 0600 takeoff. We should be back by 0730 or 0800. Why don't you meet us here then? You can listen to the debrief."

"Yes, ma'am." Lieutenant Schmid rose, gathered up his books, and left the room.

Connie turned to Greg Robinson. "For three weeks you've had volumes of information to digest, Lieutenant Robinson. You may have felt as if you were drinking out of a fire hydrant. Today you'll be flying in a T-38 and putting that information to use. You've got to be sharp and think sharp, able to handle any emergency in half time. . . ." He's wondering if he can do it, and if he can trust me to teach him. "Let's go sign out, and then we'll brief the flight."

As they passed the desk Connie and her student signed out on the board. After briefing, they went to the parachute shop, picked up helmets and parachutes, and headed for the trolley. Woody, the driver, walked out with them. He smiled and said to Lieutenant Robinson, "Don't worry, kid. She can fly. I took her out to her plane the day she soloed the T-37—watched 'em dunk her in the tank, too. She's one of our best pilots, I'll tell you that."

Connie just smiled.

Woody kept up the chatter. "Yes, sir, Lieutenant, I've been on this flight line ten years now. I was born in Texas and came out here with my wife to retire. But there's something about flying that draws a person, you know? I've watched many a young pilot come and go from this field."

They boarded the trolley and Woody started it moving, headed for the T-38 with 4919 painted on the tail. Woody knows the tension a pilot feels on the first flight with his instructor, Connie thought, even though he's never been off the ground. He loves these guys, picked up their flying vocabulary years ago. She looked at the back of his head. His gray hair was cut short like the pilots' at Willie.

The trolley stopped. "You fly straight and level, Captain Engel, and, Lieutenant, you hang in there. Someday you'll be able to do all the maneuvers the Captain's going to show you today." Woody waved a cheerful good-bye and turned the trolley back to the squadron.

After a pre-flight inspection, both pilots climbed into the plane, put on their helmets, and strapped in. Lieutenant Robinson went through the checklist to start the plane as Connie reviewed her checklist to make sure he didn't miss any item. A few minutes later, engines running, Connie called for taxi, and Lieutenant Robinson moved the plane out. Before turning onto the taxi strip, he stopped to wait for an F-5. Radio chatter indicated the area was busy.

"AWOL 52, taxi."

"Roger, AWOL 52. Taxi to runway 30, altimeter 30.02."

"Mugger, taxi with two fighters."

"T-38 on taxiway 7 maintain right side for back taxiing T-38."

"Roger."

"Aircraft on taxiway 7, use caution for emergency vehicles."

"Williams new altimeter 30.04."

Now it was their turn.

"AWOL 32, number one for takeoff."

"Roger, AWOL 32. Taxi into position and hold."

"Roger, on to hold."

"AWOL 32, winds 310 at 8 knots. Cleared for takeoff."

"AWOL 32."

Her eyes alert to every move, Connie directed Lieutenant Robinson to move into position on the runway, add power, and hold the brakes.

"O.K. That's fine," she said.

"Instruments O.K. Master caution light out."

"Let's go."

The plane raced down the runway and lifted into the air as Connie sat poised to take over the controls in a split second. No time to relax.

"O.K. Let's climb on up there. Keep your nose up."

Lieutenant Robinson was quiet as Connie helped him with instrument readings. He's wondering if he can think fast enough to initiate the proper maneuvers in time, Connie thought. "We're headed out to the area, now," she said. "I'll show you a loop, cuban eight, a couple of stalls and recoveries—let you get the feel of the airplane a little better. You're going to love this airplane, Lieutenant. By the end of the hour you'll feel much more comfortable in it. I promise."

As they approached a layer of snow-white clouds, Connie took the controls and climbed straight through them. Time enough for Greg Robinson to learn to fly in clouds, she thought. Now at 25,000 feet she rolled the supersonic jet gently to the right and caught the bright morning sun just coming over the horizon. The sun's always there, she thought. The plane smoothed out on a straight course, gliding on top of the clouds, as her fingers touched the stick. Lord, she prayed, Greg's going to fly in clear and cloudy skies. I'm going to be with him for awhile, teaching him all I know. Later, he'll be alone. Lord, more than anything else help me to show him that there's love and life beyond the clouds.